INDUSTRIAL TOURISM

The European Institute for Comparative Urban Research, EURICUR, was founded in 1988 and has its seat with Erasmus University Rotterdam. EURICUR is the heart and pulse of an extensive network of European cities and universities. EURICUR's principal objective is to stimulate fundamental international comparative research into matters that are of interest to cities. To that end, EURICUR coordinates, initiates and carries out studies of subjects of strategic value for urban management today and in the future. Through its network EURICUR has privileged access to crucial information regarding urban development in Europe and North America and to key persons at all levels, working in different public and private organizations active in metropolitan areas. EURICUR closely cooperates with the Eurocities Association, representing more than 100 large European cities.

As a scientific institution, one of EURICUR's core activities is to respond to the increasing need for information that broadens and deepens the insight into the complex process of urban development, among others by disseminating the results of its investigations by international book publications. These publications are especially valuable for city governments, supranational, national and regional authorities, chambers of commerce, real estate developers and investors, academics and students, and others with an interest in urban affairs.

EURICUR website: http://www.euricur.nl

This book is a joint publication of EURICUR and the Institute for Housing and Urban Development Studies, two members of the Erasmus Centre for Urban Management Studies. It is one of a series to be published by Ashgate under the auspices of EURICUR, Erasmus University Rotterdam. Titles in the series are:

Growth Clusters in European Metropolitan Cities
Leo van den Berg, Erik Braun and Willem van Winden

Information and Communications Technology as Potential Catalyst for Sustainable Urban Development
Leo van den Berg and Willem van Winden

Sports and City Marketing in European Cities
Leo van den Berg, Erik Braun and Alexander H.J. Otgaar

Social Challenges and Organising Capacity in Cities
Leo van den Berg, Jan van der Meer and Peter M.J. Pol

City and Enterprise
Leo van den Berg, Erik Braun and Alexander H.J. Otgaar

The Student City
Leo van den Berg and Antonio P. Russo

European Cities in the Knowledge Economy
Leo van den Berg, Peter M.J. Pol, Willem van Winden and Paulus Woets

The Safe City
Leo van den Berg, Peter M.J. Pol, Guiliano Mingardo and Carolien J.M. Spellier

E-Governance in European and South African Cities
Leo van den Berg, Andre van der Meer, Willem van Winden and Paulus Woets

National Policy Responses to Urban Challenges in Europe
Leo van den Berg, Erik Braun and Jan van der Meer

Empowering Metropolitan Regions Through New Forms of Cooperation
Alexander Otgaar, Leo van den Berg, Jan van der Meer, Carolien Speller

Industrial Tourism

Opportunities for City and Enterprise

ALEXANDER H.J. OTGAAR

LEO VAN DEN BERG

CHRISTIAN BERGER

RACHEL XIANG FENG

European Institute for Comparative Urban Research
Institute for Housing and Urban Development Studies
Erasmus University Rotterdam
The Netherlands

In cooperation with:
Sino-Europe Comparative Urban Research Center
Shanghai Normal University, China

Routledge
Taylor & Francis Group
LONDON AND NEW YORK

First published in paperback 2024

First published 2010 by Ashgate Publishing

Published 2016 by Routledge
4 Park Square, Milton Park, Abingdon, Oxon OX14 4RN

and by Routledge
605 Third Avenue, New York, NY 10158

Routledge is an imprint of the Taylor & Francis Group, an informa business

Publisher's Note
The publisher has gone to great lengths to ensure the quality of this reprint but points out that some imperfections in the original copies may be apparent.

British Library Cataloguing in Publication Data
Industrial tourism: opportunities for city and enterprise.
-- (EURICUR series)
1. City promotion--Case studies. 2. Tourism--Case studies. 3. Corporations--Public relations--Case studies.
4. Public-private sector cooperation--Case studies.
I. Series II. Otgaar, Alexander H. J. III. European Institute for Comparative Urban Research.
659.2'930776-dc22

Library of Congress Cataloging-in-Publication Data
Industrial tourism: opportunities for city and enterprise / by Alexander H.J. Otgaar ... [et al.].
p. cm. -- (EURICUR series)
Includes index.
ISBN 978-1-4094-0220-6 (hardback) -- ISBN 978-1-4094-0221-3 1. Industrial tours. 2. Industrial relations. 3. Tourism--Industrial capacity. 4. Tourism--Political aspects. 5. Tourism--Government policy. I. Otgaar, Alexander H. J.
T49.5.I534 2010
910.68--dc22

2010007200

ISBN: 978-1-4094-0220-6 (hbk)
ISBN: 978-1-03-292569-1 (pbk)
ISBN: 978-1-315-58843-8 (ebk)

DOI: 10.4324/9781315588438

Contents

List of Figures

List of Tables

Acknowledgements

This book is the result of an international comparative study on industrial tourism development in cities. It provides insight in the conditions under which industrial tourism can be successfully developed, resulting in concrete recommendations for cities and companies with ambitions in this field. The book demonstrates how industrial tourism can become an instrument for improving the image of a city and its industries, stimulating the competitiveness of cities and regions (as tourist destination and business location) and enhancing the relation between companies and society.

The international study was commissioned by three Rotterdam-based organisations: Rotterdam Marketing, the Rotterdam Development Corporation (OBR) and the Academic Centre TransPORT. It has been carried out by the European Institute for Comparative Urban Research (Euricur) and the Institute for Housing and Urban Development Studies (IHS), both based at Erasmus University Rotterdam, in cooperation with the Sino-Europe Comparative Urban Research Center (Securc), based at Shanghai Normal University.

The empirical study consists of six case studies: Wolfsburg (Germany), Cologne (Germany), Pays de la Loire (France), Turin (Italy), Shanghai (China) and Rotterdam (The Netherlands). Five of the six case studies have been organised in cooperation with local experts: Walter Fink (Wolfsburg), Julie Leblanc and Sophie Hermange (Pays de la Loire), Irma de Winter (Rotterdam), Fabrizia Pedroletti (Turin) and Prof. Gao Jun (Shanghai). With their help more than 60 discussion partners have been identified representing public and private organisations involved in industrial tourism (see the list of discussion partners at the end of each case study). We would like to thank all the abovementioned organisations and people for their valuable contributions to the realisation of this book.

Finally we would also like to thank Erik Braun, Erwin van Tuijl, Lenneke Wester and Ankimon Vernède (all members of Euricur) for their support.

Alexander Otgaar
Leo van den Berg
Christian Berger
Rachel Xiang Feng

Chapter 1
Industrial Tourism Development in Cities: Towards a Framework

1 Introduction

Industrial tourism is a type of tourism which involves visits to operational companies and industrial heritage; it offers visitors an experience with regard to products, production processes, applications, and historical backgrounds. More than 100 years ago public and private organisations already started to organise industrial tourism visits. Historic examples include visits to French wineries and chocolate factories, to Greek and Maltese lace-producers and to Dutch cheese producers. In Paris, tourists and residents had the opportunity to choose among various tours back in 1900. The tours on offer ranged from visits of textile-factories to tours of mortuaries and slaughterhouses. Furthermore, also a visit to a tobacco factory, the stock exchange, the parliament and the court of justice were possible. There is also evidence from the United States that industrial tourism has actually a long history; a good example is the Jack Daniel's distillery in Tennessee which began to offer factory tours back in 1866, the year when the company started operating (Frew, 2000).

Industrial tourism offers opportunities for individual firms, but also for cities. Particularly for cities with a considerable industrial base, industrial tourism comprises interesting possibilities to strengthen the economic structure (direct and indirect employment) and to increase the supply of tourist products. For such cities industrial tourism is a potential growth sector that matches with their identity: the sector offers opportunities to strengthen their distinctiveness and image, notably by building on their existing assets. Many cities, however, have the assets to develop industrial tourism but fail to capitalise on them. In most cases industrial tourism is only restricted to low-profile company visits, without any clear vision of the municipality or firm on the strategic value of it, not to mention coordinated product development.

The development of industrial tourism largely depends on the ability of firms to cooperate. On the one hand, we observe that many firms consider company tours an obligation with more costs than revenues. Many companies are afraid of leaking confidential business information, and also security often constitutes a problem. On the other hand, more and more companies also recognise the opportunities of industrial tourism: industrial tourism can help to market their product, to secure their license to operate (reputation), and to remain attractive as employer. From the observation that industrial tourism offers opportunities for city and enterprise, and the fact that in many cities these potentials are not utilised to the full, this book addresses the question: under what circumstances can industrial tourism make a strategic contribution to the competitiveness of city and enterprise?

This book provides insight into the opportunities of industrial tourism development, the barriers that obstruct the utilisation of potentials, and the possibilities to reduce these barriers. In the present chapter we develop a conceptual framework that identifies the conditions for successful industrial tourism development, based on literature study. In the next chapters, the framework will be assessed by applying it to six case studies: Wolfsburg (Germany), Cologne (Germany), Pays de la Loire (France), Turin (Italy), Shanghai (China) and Rotterdam (The Netherlands). In the final chapter we compare the different case studies and formulate general conclusions and recommendations. For the case studies, information has been collected by means of desk research and semi-structured interviews with representatives of stakeholders in industrial tourism.

In the present chapter we first define the term industrial tourism (section two) followed by an analysis of the demand-side (section three) and supply side (sections four and five) conditions for a successful development of company visits. The sixth section analyses the conditions under which the interests between the public and private sector concerning industrial tourism converge. The final section presents a general frame of analysis to be applied to the case studies, formulating four conditions for developing industrial tourism successfully.

2 Defining industrial tourism

Industrial tourism is a relatively unexplored research theme, with a lack of clear concepts and definitions. First, the term industrial tourism has different meanings. Particularly in the US, some use the term industrial tourism when they talk about a commercial approach to tourism, maximising the return on investment (for example Wemhaner, 1976). From this point of view, industrial tourism implies the marketing of tourism as a commodity to be sold for mass consumption (Abbey, 1968). Second, other terms are used to describe the same phenomenon. In some cases, this seems only a difference in language: Industrietourismus (German), tourisme industriel (French), industrieel toerisme (Dutch), turismo industrial (Spanish), turisme industriale (Italian), etc. However, it appears that in some countries industrial tourism is mainly about visiting operational firms, while in other countries it refers to visiting industrial heritage (non-operational firms), or comprehends both meanings. In Germany and France, for instance, many people associate industrial tourism with industrial heritage, and not with visiting operational firms. The terms *Werkstourismus* (factory tourism) or *Betriebsbesichtigungen* (company visits) are more widespread in Germany, while also the French prefer to use the more neutral *visites d'enterprises* (company visits). In the Anglo-Saxon literature it is common to make a distinction between industrial tourism (visiting operational firms) and industrial heritage tourism. In other countries, the term industrial tourism is not very common. Instead, people use the more neutral terms company visits or factory tours. The main advantage of using these labels is that they are not directly associated with tourism and (leisure) tourists, thus allowing for various motives to visit a company (leisure, education, knowledge exchange, etc.). Others talk about technical visits in case of visits with an educational or professional motive. Also the name industrial attractions is widely used. Especially in Japan the term industrial sightseeing is used as a synonym for visits to operational companies.

Also among scholars we observe some confusion when it comes to defining industrial tourism. Frew (2000: 20) states in her empirical study of industrial tourism attractions that 'industrial tourism involves visits by tourists to operational industrial sites where the core activity of the site is non-tourism oriented'. Visits to non-operational firms are obviously excluded from this definition. Soyez defined industrial tourism as movements caused by the appeal of former or operational

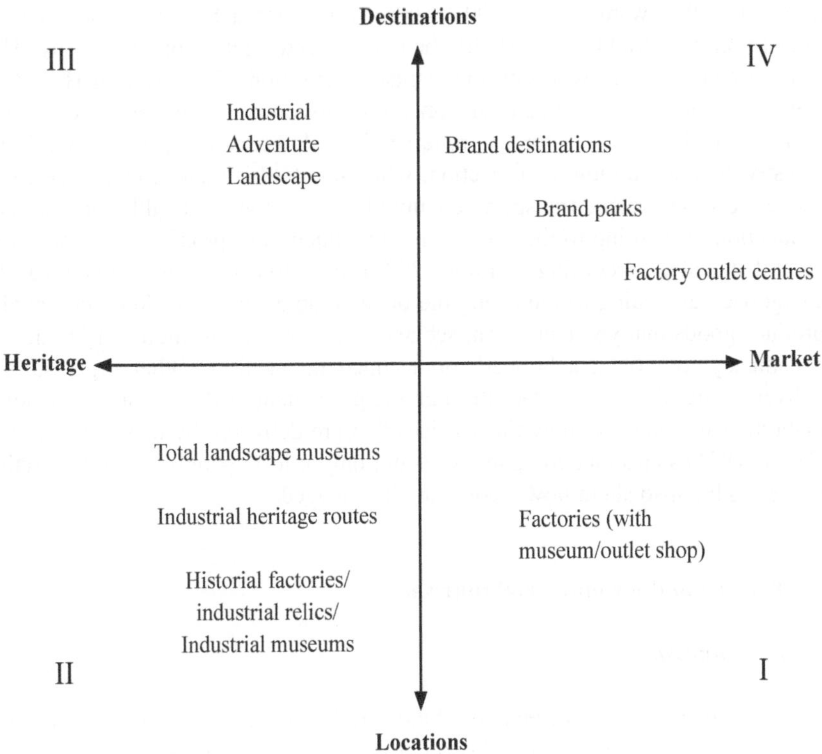

Figure 1.1 A two-dimensional model for categorising industrial tourism products

Source: Industrial tourism destination management in Germany: A critical appraisal of representation practices (Li and Soyez, 2006), with permission.

industries on external visitors (1993). Marcon et al. (2000) consider visits to operational firms as one of the three categories of economic discoveries, the other two being science tourism (including visits to museums and science centres) and industrial heritage tourism (*le tourism de patrimoine industriel*). Li and Soyez (2006) present a two-dimensional model for categorising industrial tourism products, with the horizontal axis indicating the difference between heritage and market (operational firms) and the vertical axis going from locations to destinations. Company visits combine a location with the market, whereas brand parks are examples of destinations in the market.

We also observe some disagreement concerning the 'industries' that are to be visited. Yale (1991) defines industrial tourism as the presentation of contemporary manufacturing processes, thus leaving out visits to service companies. In a similar way, Swarbrooke (1995: 51) talks about factory tourism defined as 'visits to working factories to see industry in action, mainly manufacturing industry'. MacCannel (1976), however, states that all firms in society can become a tourist attraction if they want. Frew (2000) uses a similar approach, arguing that the term industry in industrial tourism should be read as sector not as manufacturing. This last view goes along well with the standard definition of industry of Hill et al. (1975): this definition of industry not only refers to companies that produce tangible goods, but also includes for example services. Industry comprises agriculture, forestry, fishing, mining, construction, wholesale trade, finance, insurances, real estate, services, retail, transport, communications, energy, health services and production. According to these scholars, manufacturing (production) is only one part of industry. According to Frew (2000), industrial tourism products can be categorised according to their tangible or intangible character. Some companies produce goods that you can touch, see and sometimes even smell or taste during a company tour (e.g., a brewery or a chocolate factory). Other organisations deliver intangible goods: a tour through the parliament or through an educational institution will mainly show the services that are delivered by these institutions. Carter (1991) states that company tours not only inform people about production processes but also about how a company is managed.

3 The demand for industrial tourism

Trends in tourism

Before we discuss the development of industrial tourism (the supply side), we first need to assess the demand for industrial tourism. Many scholars and professionals claim that this demand will grow, but on what grounds? They mainly refer to trends in tourism.

The first argument is that the tourism industry is growing, due to a globalising world that allows more and more people to travel. Between 1990 and 2006 the number of tourist arrivals worldwide increased with more than 100 per cent (from

436 to 846 million) (UNWTO, 2007). Considering the countries tourists come from (outbound statistics), it appears that particularly Asia, the Middle East and Africa show high annual growth rates (6.4 per cent; 10.3 per cent; 7.1 per cent, average: 3.6 per cent). The World Tourism Organisation predicts a further growth of the tourist industry to more than 1.5 billion arrivals in 2020.

The countries of Western Europe belong to the most mature origin markets for tourists worldwide: a broad scope of people from all social classes engage in tourism. Increasing real incomes and leisure time, coupled with declining costs of travel and an increasing car ownership rate, lead to a growing demand for recreation and holidays (Williams and Shaw, 1992), which is what we could observe in the Europe over the past decades. According to Williams and Shaw (1992) this is the reason why tourism is regarded commonly as one of the economic sectors which has realistic potential for growth beyond the short term. A further increase of income and leisure also went along with a growth of second and third holidays, including a trend towards short breaks and especially towards urban tourism. This growing demand for tourism is distributed between domestic and foreign destinations, thus we can observe a growing demand both for the domestic and the foreign market.

As we have seen above, destinations that are far away show high growth rates. This indicates that the tourists themselves have become more mature and demanding: after already having visited rather traditional tourist attractions/places like London, Paris or Venice, they long for more exotic destinations far away. Analogically, this observation also implies that 'the well travelled tourist' is already saturated by 'classic tourist cities' and standard tourist attractions like churches, museums and so on. This offers new opportunities for urban tourism in second cities that are not (yet) on the shortlist of the 'must see attractions'. Van der Borg (1994) notes that especially the Netherlands, Germany, the UK and Denmark can be characterised as very mature outgoing tourism markets; this is also reflected in the destinations tourists from these countries tend to visit. Tourist flows from these highly mature outgoing markets are much more diversified than those in less mature markets for outgoing tourists, like Spain: Spanish tourists still tend to visit the classic tourism cities (Van der Borg, 1994). But of course, the matureness of outgoing tourism markets is subject to constant change, and as more and more (European) countries enter the mature stage, also the demand for visits to secondary cities and attractions will increase.

Especially cities that have an alternative blend of attractions on offer, like industrial tourism attractions, can be interesting for experienced tourists that have already seen most of the must see attractions. Especially foreign countries with a lot of mature and demanding tourists are interesting target markets for industrial tourism. Taking into account the domestic visitor market for industrial tourism, we can assume that cities can consider their location within a mature outgoing tourist market (like the Netherlands or Germany) as an advantage; a higher number of mature tourists potentially comes along with a higher domestic demand for industrial tourism. Industrial tourism attractions that are located in regions with a high population density, like the Dutch Randstad or the German Ruhr area may have an additional advantage in terms of domestic tourism: a high absolute number of local and regional visitors can be expected.

Industrial tourism is not only a relevant market for second cities, but also an interesting addition for rather traditional tourist cities that are already successfully attracting tourists; visitors ask for a variation of the standard urban tourist programme, and industrial tourism could be an interesting tool to 'add some spice' to the existing tourist attractions, and hence create more appeal for second time visitors. Urban tourism consists of two very distinctive main strands: on the one hand cultural tourism and on the other hand business/conference tourism. According to Murphy (1982) cultural tourism is very much linked to learning. Taking a wider view of the first strand of urban tourism, this includes cultural and learning experiences, both of which can be provided by industrial tourism.

The importance of learning is also underlined by Martin and Mason (1993), who predict that visitors will become more selective in the future, concerning the choice of visitor attractions. In the future, visitors will pay more attention to purposeful visits, thus visiting places and attractions that are related to their specific interests. The share of tourists that visit an attraction just for the sake of spending their free time – and without any further purpose – will become smaller and smaller in the future. What Martin and Mason actually predict is a shift in the popularity of attractions: a shift from passive attractions to active attractions, and a shift from pleasure to education. This shift implies that traditional museums or visits of historical houses will diminish, while for example visitor numbers of interactive science parks and museums will increase. What could this mean for industrial tourism? Industrial tourism is an educational experience and is well placed on the market in this respect, but its relative position towards other tourist attractions could be even enhanced further if the visitors can be involved actively in (certain steps of) the production process, for example by being allowed to operate a machine.

Evidence of increasing demand

Regrettably, the World Tourism Organisation does not provide any insight in the demand for company visits or industrial tourism. To assess the demand for company visits, we depend on various national and local sources. For instance, in New South Wales (Australia) more than 10 per cent of the tourists visited an industrial tourism attraction in 2004. In the same year, almost eight million people visited an industrial site in France. In the UK, industrial tourism could attract around five million visitors per year in 1989. One year later 294 UK companies were open for visitors; these companies accounted for 6 per cent of all British tourist attractions (!). Frew (2000) provides an overview of some industrial tourism attractions (in the US, the UK and Japan) and their visitor numbers, giving an idea about the dimension of industrial tourism (see Table 1.1).

Table 1.1 A selection of visitor attractions in the UK, Japan and the US[1]

Industrial Tourism Attraction	Visitors per year
Cadbury World (UK, chocolate factory)	400,000
Toyota Motor Corporation (Japan)	300,000
Ben and Jerry's Homemade (USA, Ice-cream Production)	170,000
Wedgwood (USA, Pottery Factory)	100,000
British Nuclear Fuels (USA, Nuclear site clean-ups)	100,000
WR Outhwaite and Son (USA, Ropemakers)	75,000–100,000
Cumberland Pencils (USA)	64,000

Source: Frew (2000)

These numbers suggest that there is demand for industrial tourism from the broad public, and not only from a few special interest groups. Furthermore we can see that not only attractive consumer products like chocolate or cars can attract huge visitor numbers. Immense visitor interest is also directed towards 'business to business products' like nuclear fuels or ropes. Especially for the manufacturing sector, the character of the goods (business to business or consumer goods) seems to be secondary – both Harris (1989) and Prentice (1993) note that there is an increasing public interest in the manufacturing sector in general. In the United States and in Europe, the shift from a manufacturing to a knowledge economy has dislocated people spatially and culturally from the manufacturing sector. Especially the younger generation has very little knowledge about industrial work; according to Harris and Prentice, this lack of knowledge naturally leads to an increasing curiosity about the topic, and thus to a demand for company visits in the manufacturing sector. On the other hand, older people who had still been employed in the manufacturing sector years ago, may the feel the need to 'return to their roots', thus a demand for industrial tourism can be detected in all age groups.

Mader (2003) collected evidence of rising demand of company visits in the German Ruhr area. His study is especially interesting, because the surveyed companies do not engage in any promotional activities to enhance their prominence and visitor numbers. Nevertheless, six of the seven companies that were surveyed showed an increasing number of tourists in the ten years preceding the survey (1993–2003). For example the Chemical Park Marl managed to increase its visitor number from 915 in the year 1999 to 2859 in the year 2002, whereby the park offers constantly six tours per week in summer and three tours per week in the winter period (Mader, 2003). As this example already suggests, the demand of industrial tourism depends very much on the period of the year and as well on the target group. Visits of professionals are usually concentrated in the months

1 The visitors numbers have been collected from 1984–1993 and thus do not refer to the same year.

February until June and September until November, during the week. On the other hand, leisure visits rather take place during the school holiday period and at the weekend (Mader, 2003).

Visitors go to a city, not because of one tourist attraction, but because of the whole package that is offered. In line with this philosophy, Van der Borg et al. (1995) make a distinction between primary tourist products (museums, amusement parks, cathedrals, etc.) and secondary tourist products (hotels, restaurants, shops), emphasizing the need for a well integrated tourist product. Baumann (1999) showed in a survey among a number of companies, that 95 per cent of the industrial tourists combined their visit with other primary or secondary tourist attractions in the region, like a city tour, the visit of a theatre etc. This indicates that there could be also a demand for tour packages, involving more than one (industrial) tourism attraction in one tour. The idea behind this is that a city can be seen as a shopping mall in which customers look for various shops.

What makes industrial tourism attractive?

According to a survey that was carried out in 1992 in the UK, the main reason why people visit companies are (1) to see and experience the production process and (2) to buy goods (Frew, 2000). Mader (2003) suggests that the touristic potential of companies is mainly dependant on the kind of product that is produced. He suggests that especially consumer goods are interesting for visitors, while services and intermediate goods are relatively unattractive for the leisure tourist. Interestingly, Mader (2003) introduces a model that allows the classification of industrial tourism products. He introduces eight different categories:

- Goods with a symbolic character for the region: coal and energy for the German Ruhr area, the port and everything related to it for Rotterdam, cheese production for Gouda, bananas and coffee for Guatemala, and so on.
- Branded goods: for example cars, sports goods and beer.
- Consumer goods: for example lemonade, beer, chocolate, newspaper and toothpaste.
- Goods of everyday life: for example cars, expensive watches, diamonds and perfume.
- Luxury goods: for example cars, watches and jewellery.
- Technologically demanding goods: for example cars, computers, mobile phones and compact discs.
- Special interest goods: for example energy, water, television, public facilities, food and theatre (backstage).
- Demanding handicraft goods: for example glass, chinaware and blacksmith products.

According to Mader, these categories operate cumulatively: thus, a product has more touristic appeal (and can create more demand for company tours), if it belongs to several of these categories. Observing the list of categories above he notes that especially a car belong to a lot of the listed categories: it is a branded good, a luxury good, a key good of everyday life and it is a technologically demanding good. Moreover, big car factories also very often dominate a region and hence also reflect their characteristics. This could probably also be the reason why the car industry is a forerunner when it comes to company visits. The attractiveness of the product and hence the demand for company tours is so high, that car manufacturers were 'naturally driven' to develop company tours. The data from other authors (for example Gelbert, 1994) suggests that also the drink and food industry belongs to a segment that is highly attractive for visitors. A disproportional part of industrial tourism attractions worldwide comprises companies belonging to this category.

1 The supply of industrial tourism from a private- sector perspective

Benefits

By receiving visitors, companies may gain access to numerous benefit. A 1992 survey of company visits in the UK brought to light that the most important reasons why companies engage in industrial tourism are (1) to improve their image, (2) to improve the working morale and (3) to reap an extra income (Frew, 2000). These and other benefits will be discussed below.

First of all, industrial tourism can make a substantial contribution to the marketing of products and the enhancement of brands and company image. An increase in buyer loyalty, in sales and market share can be the consequence. Involvement theory suggests that consumers who have attended the production of a good become more brand loyal. This is due to the increased familiarity with the product, their first hand interaction with the employees and their first hand observation of quality assurance procedures. According to Mitchell and Orwig (2002) more and more brand managers are recognising the potential of industrial tourism to strengthen the bonds with (potential) customers, and increase brand loyalty. An illustrative example is the reaction of the director of Celestial Seasonings – an American tea producer – on the huge success of their plant tour: 'One visit has more value than one hundred commercials when guests relate their positive experiences to friends and family.' (Brumberg and Axelrod, 1995). In other words: industrial tourism offers plenty of possibilities to make visitors enthusiastic about the firm and its products, indirectly also reaching other stakeholders through word of mouth advertisement. Industrial tourism is an especially promising marketing instrument, as the target group that can be reached (leisure visitors and professionals) is typically highly educated, with a high share of people that hold an influential position of some kind. Even considering the costs involved in industrial tourism (like the salary of the guide), it can be a rather attractive marketing instrument. One of the interviewees of Mader (2003) noted that the yearly salary of a guide costs only about as much as a double sided advertisement in the German magazine *Spiegel*. Industrial tourism can be also an interesting marketing instrument for companies that produce intermediate goods or services. For these kind of companies, industrial tourism can be an instrument to influence individuals in other companies (their business to business customers, their suppliers but also for example their sales agents and shareholders) and increase their loyalty. Groups of students are also an attractive target group for such companies. Students with an interest in the topic are very likely to work in a related area in the future; it is quite likely that at least some of the youngsters will choose to work for a supplier or customer of one's own company in the future. If these students hold a good opinion about one's enterprise, the probability that they will opt to purchase the company's products is high, once they reach a corporate position with decision power. But not only students and employees of potential business to business customers are relevant target groups. Every consumer in general could be relevant: companies that produce intermediary products can also showcase the contribution of their product or service to a consumer product. Indirectly, an intermediary company can thus benefit from the increasing demand for their intermediate product which is caused by a stimulation of sales of a consumer product that contains their inputs.

Secondly, industrial tourism can support the daily production process. Company tours potentially contribute to productivity gains because employees become proud of the firm they work for (following the positive attention of other people) which ultimately increases their working morale. According to Mitchell and Orwig (2002) everybody who interacts with a company (also buyers, suppliers, managers, employees, and so on) benefit from a close look on the production process; so productivity gains could also be achieved through better knowledge of the process and an the resulting improved co-ordination with customers and suppliers.

Thirdly, industrial tourism might create opportunities to generate turnover, for instance by means of selling regular products or souvenirs, or by catering. Direct income can also derive from the sale of entrance tickets. These benefits are particularly relevant if companies consider industrial tourism as a (commercial) profit centre. Moreover, industrial tourism can offer an opportunity to sell slightly damaged products – or products with a damaged packaging – on the spot (slightly damaged products usually occur in any production process). This can reduce waste disposal costs and can even generate extra income for the company (Henly, 1988).

Fourthly, industrial tourism responds to the need for corporate social responsibility and civic engagement. Company tours can be part of strategy that aims to secure the license to operate. By receiving visitors, companies can show how they treat their personnel, the environment and the communities in which they operate in general. This strategy is particularly interesting for companies that run the risk of negative publicity and reputation damage, like for example the chemical sector or the energy sector. Industrial tourism can be used as a way to build a relationship with the local residents – convincing them of the positive aspects of their company and allaying fears. Company visits can also be an instrument against lay offs and to employ people further in bad economic times. Especially in regions that face a decline of traditional industries, industrial tourism can be used to shift people within a company from manufacturing jobs to jobs created through industrial tourism. This is one way how companies can show their corporate social responsibility.

Fifthly, industrial tourism can help to attract employees. Company tours help to make potential employees – notably students and scholars – interested in the production process, and positive about the company. Particularly when labour markets are tight company tours become an important instrument in the human resources strategy. Furthermore, investments in industrial tourism help to create an attractive city – in terms of facilities and image – and thus, an attractive place to live for the higher educated.

Sixthly, industrial tourism can be seen as a way to get in touch with the customer and to be close to the customer's wishes. Company visits can thus serve as a kind of market investigation instrument, a continuous process that feeds back information and allows for a quick adjustment to changing customer wishes. Especially for companies that are separated from their final customers through intermediaries, company tours are an instrument to get in direct contact with the customer.

Costs

The costs of industrial tourism cannot be seen in an isolated way. Of course, companies need to assess these costs with regard to the benefits that can be reached, in order to assess if industrial tourism makes sense for them. Most of the benefits mentioned above can be collected on the medium and long run only, which means that companies that only think about the short run will probably be very difficult to convince to open their doors for company tours. Thus, in the first place the mindset and attitude of an entrepreneur is of major importance for the potential to realise company tours. If an entrepreneur does not think about the medium and long run, industrial tourism is rather out of question. For companies that produce a business to business product, the marketing advantages of industrial tourism are generally less obvious. There are, however, several other possible reasons why these companies could have an interest in industrial tourism. The question is if companies are actually aware of these benefits.

Apart from a lack of awareness, most of the barriers of industrial tourism can be translated as the costs of industrial tourism: of course, it costs money to employ a guide or to create visitor facilities. Small companies might not be able to offer company visits on a regular basis, simply because of a scarceness of resources (Mader, 2003). If such companies are supposed to open their doors, they might need some assistance from the public sector in order to be able to organise company visits. But costs do not only refer to initial investments and running costs, they also refer to the potential risks that might occur. there are a number of constraints that can act as barriers for industrial tourism. Nevertheless, as we will see hereafter the big majority of these constraints can be tackled relatively easily, provided that the companies show the will to do so.

Companies that are very active in industrial tourism frequently face limits of visitor capacity. In many cases, however, it is rather the unwillingness of the company – and not this capacity limit – that restricts the number of visitors. The number of visitors that is admitted within a specific timeframe and in certain areas can often be increased, for example by professionalising the tours or by finding innovative ways to allow for a higher carrying capacity. But obviously, there will be a maximum number of visitors that can tour a company during a defined timeframe, and companies need to ensure not to exceed this limit. If this restriction is violated, the company may encounter negative consequences, such as a delay of the production process, which ultimately means additional costs for the company. Moreover, some departments of certain companies need to be off limits for visitors, simply because the presence of visitors would disturb the production process (for example in a paint shop or in a laboratory) or because it is dangerous for visitors to be there. But even in such cases innovative solutions can be found, like for example a glass tunnel that runs through a laboratory.

Industrial tourism may also open the door for theft and industrial spying. In practice, however, theft from visitors is not a major concern of companies and also the risk of spying can be easily reduced, simply by keeping sensitive areas of companies (like laboratories or research departments) closed from spectators, or by suspending tours of certain areas during sensitive periods, like during the introduction of a new model. Furthermore, the employees themselves can be opponents of guided tours in their company; sometimes, employees may feel uneasy when visitors are around and watch them, which can have negative consequences on the production. However, in most cases companies report exactly the opposite – namely that their employees feel more motivated and proud of their work – because of the visitors.

Another difficulty associated with company tours, is the often very seasonal character of tourism. This is not so much of a problem if the tours are only offered every now and then when somebody asks for it. But this may become a challenge, when company visits are offered on a regular basis. What to do with the guides that are not needed any more in the off season? We can preliminary reckon that there are two main ways to deal with this: on the one hand, companies can employ guides only on a flexible basis, or they can – as Mader (2003) suggests – dip into the pool of their already retired employees as tour guides; the latter could be a rather cheap and flexible way to organise company visits. If a company chooses for a permanent employment of guides, the company might have to act more proactively and try to approach schools and other organisations especially in the low season of industrial tourism. They could for example align their visits more to the educational programmes of universities. Other challenges for companies are the availability of sufficiently big parking areas, sanitary facilities and a kind of front room where they can receive the visitors (Frew, 2000). However, all of these challenges can be solved if the company tackles them in a proactive, anticipatory and professional way.

Mader (2003) also mentions another reason that could act as a barrier for industrial tourism: for some companies opening their production process to the public could be counterproductive for their desired image. For example a company that is producing sausages may not want to display all of its activities to the public, as people might feel disgust of the product after having observed all the steps of the production process. Another reason not to open one's company could be that the product lacks acceptance in wide parts of society. This could for example concern producers of mink coats or cigarettes. And indeed, the cigarette factory R.J. Reynolds decided to stop its (very popular) tours because of rising anti smoking sentiments in the US (Mitchell and Orwig, 2002). On the other hand one could argue that a negative image of one's product can be an extremely good reason to start offering a company tour; in order to show that everything is done according to regulations and to correct wrong perceptions. According to Frew (2000), companies also may choose deliberately not to open their doors, simply because they think that there is nothing interesting to see. An example are factories where the whole production process occurs within machines that are completely closed towards the outside. Nevertheless, such kind of companies could still be interesting for professional visitors.

Obviously, not all companies manage to deal with the challenges mentioned in this section in an adequate way. Though the entire number of companies that organise company tours is increasing, in the 1990s some companies in the United States actually decided to reduce or stop their company tours, and to shift the focus on – easier to manage – company museums instead. Also in France some industrial attractions had to close, due to an external decision: a national security plan (Vigipirate) has for example put an end to tours of nuclear plants. The disadvantage of replacing company tours by company museums is that visitors cannot experience real work any more. According to Frew (2000) this is a serious threat for the attractiveness of the visits, as authenticity is lost. A less attractive visit may also mean that many of the benefits of industrial tourism cannot be fully reaped any more.

2 Company visits from a public sector perspective

On the one hand cities can act as a facilitator, encouraging private companies to open their doors for visitors. Secondly, cities can also be owners of companies (municipal enterprises) themselves and can influence the supply of industrial tourism directly (for example tours of public companies, tours of the city hall and so on). In this book, however, we focus on the first role. The question is: should cities have an explicit strategy for industrial tourism and should they encourage companies to open their doors for visitors? What are the benefits for cities, in terms of income and image?

Income effects

The direct effects of tourism can be measured by the share of tourism in regional income: if an increase of this share can be observed, the city has benefited from tourist activities. However, also multiplier effects have to be taken into account, to be able to grasp the full benefits for a city or region (Williams and Shaw, 1992). It is very difficult to determine the extent of these multiplier effects, but just to give a hint about their possible dimension we want to mention that Archer (1977) had calculated that the tourism income multiplier is between 1.68 and 1.78 for the UK. Williams and Shaw (1992) point out that multiplier effects are greatest in large tourist economies and in specialised tourist economies, because these economies are more able to support linked specialised services. For second cities without a strong tourism economy, this indicates that it could be beneficial for the city to promote industrial tourism as a spearhead policy, rather than as an insignificant niche. Specialisation could lead to the emergence of locally specialised services (like for example industrial tourism consulting or tour agencies specialised on industrial tourism), which could develop a reach far beyond the region in question.

As we already touched upon above, tourism as such is commonly regarded as one of the economic sectors which has realistic potential for growth beyond the short term; due to rising incomes and more free time, the future potential of tourism is considered to be high (Williams and Shaw, 1992); moreover, the potential of urban tourism in particular is very high.

Baumann (1999) indicates that industrial tourism is especially interesting for cities, considering that it is a form of tourism that aims to exploit existing attractions to the full: it does not use up additional space in a city. The use of existing attractions is also related to another advantage of industrial tourism: it can potentially be implemented within a rather short time frame, probably even more quickly than other kinds of tourism. Williams and Shaw (1992) note that tourism as such can be developed in a rather short time-span, and that relatively moderate levels of investment can lead to rapid impacts on the regional economy.

In terms of employment the tourism sector holds a lot of potential, especially for traditional industrial cities that very often suffer from a rather low educated population and a lot of people living on social benefits. Direct employment in tourism (bars, hotels, restaurants and so on.) consists mainly of unskilled employment, which provides a possibility to employ a bigger part of the unskilled part of a city's population; this may in the end alleviate social problems in industrial cities. But tourism as such can indirectly also create higher skilled employment, for example in supply industries and producer services. In the special case of industrial tourism, we can expect that a higher proportion of skilled jobs may be created, as for example especially professional company guides and marketing and communication experts are needed for industrial tourism.

Image effects

As we wrote in our introduction visits to factories have been taking place for quite a long time. Nevertheless, the involvement of the public sector as a driving force behind industrial tourism, and the idea that industrial tourism can help to change the image of a place, is a rather new phenomenon. Industrial activities have had a very negative connotation in the past, and cities that accommodated a lot of these activities were generally considered to be dull and dirty places. However, things have changed a lot in Europe over the past few decades: polluting industries have either moved away to cheaper locations abroad, or have been forced by stricter environmental rules to cut back on emissions. Though things have improved a lot in industrial cities, the image of these places still suffers from the past. Changing an image is a long term process and cannot be done overnight. Cities should not have too many illusions about this process: it will be highly difficult to change the image of a city from a dull industrial place to a second Paris. The past always sticks to a city to a certain extent, and rather then denying it, one can try to turn a prevalent negative image into a strength. Industry is not only associated with pollution, it is also associated with economic strength and prosperity. Cities are the powerhouses of the economy, and a city that is known for industry should not hide its companies, but rather show them with pride to the public. Even highly successful tourist cities, such as Munich and Amsterdam, want to be associated with industry nowadays and have recently launched programs on industrial tourism. So why should traditional industrial cities deny their character, and miss a great opportunity?

The reason why cities are still hesitant to use their industries as an asset in image building and tourism is well known. Schmidt (1988) emphasises that the direct link between industry and tourism is a rather new phenomenon, as these two terms have been rather seen as a contrast in the past. Industry was typically associated with pollution, noise, hecticness, work, and greyness. On the other hand tourism evoked rather different connotations, such as intact environment, quietness, relaxation, free time and so on. These images do of course not correspond to reality anymore: both industrial activities and the tourism behaviour of people have changed quite a bit. City trips are becoming more and more important, and it is not quietness and relaxation that tourists seek when visiting cities: rather the opposite, they want to be at the places where things are happening.

If promoted in a clever way, industrial activity and vibrant cities can go hand in hand in the perception of visitors. Not so many regions are endowed with a substantial number of industrial attractions open to public; a tourism strategy that puts industrial tourism in the centre – rather than on the side – may help a region to reach a competitive advantage over other regions. As Mader (2003) indicates, this could be an especially interesting strategy for regions that do not have yet a characteristic tourism profile and that lack other tourist attractions.

A high local and regional demand for industrial tourism is important, if we consider the fact that a critical mass of people is needed to strengthen or change the image of a place, whereby residents of the area play a key role in this process. Industrial tourism can help to boost the pride of residents about their city and can help to change outdated and wrong images. Local people are ambassadors for their cities and if the residents have a positive picture of their city, this is the best way towards strengthening or changing the image of a place.

3 Converging interests?

As we discussed in the previous section both cities and companies have a potential interest to be engaged in industrial tourism. Now the question is under which conditions these interests converge: how can industrial tourism be beneficial for both sides at the same time?

We have seen that industrial tourism attractions can be developed within a relatively short timeframe and that moderate levels of investments can lead to rapid impacts on the regional economy. This could mean that cities can reap the benefits of industrial tourism activities in the rather short run. In this sense, cities do not even have to be aware of long term benefits to be able to grasp some of the benefits of industrial tourism. In contrary, we expect that most companies need to have a long term vision in order to be engaged in industrial tourism. Nevertheless, we expect that also cities can achieve extra benefits if they have a long term view of industrial tourism; such a situation would lead to converging interests with companies and could provide a lot of additional opportunities.

Another crucial question is whether the tourist industry is able to utilise local resources. Several scholars view this as a condition in order to be able to reap the benefits of tourism to the full. The more local capital, labour and food supplies are utilised for (industrial) tourism, the better for the local economy (Williams and Shaw, 1992). In this sense, industrial tourism might have a significant advantage over 'conventional' tourism: visitors get in contact with the products of local companies, and if you can encourage them to consume the local products they have experienced during an industrial tourism visit, this will be also good for the local economy in the end. Moreover, this could mean that industrial tourism in locally owned and controlled companies might be more beneficial for cities (from an income-dominated point of view), when compared to industrial tourism in international companies or foreign companies that have a plant/subsidiary in the city. In the latter case, the full benefit of industrial tourism might not be remained in the region, but could be partially transferred to the place where the headquarters or most of the shareholders of are located. Nevertheless, if we take into account image aspects as well, there is more convergence of interest between the city and multinational corporations that have their headquarters elsewhere. We assume that the image of a city can benefit greatly from a greater visibility of well known multinational companies (for example, through industrial tourism). And vice versa these companies may as well have an interest in a positive image of the city; a good image is crucial to attract the creative class to a place and to recruit the best brains.

The choice of target groups may be the key issue to reach a situation in which the interests of companies and cities are converging. If we consider image effects, we can expect that a city will welcome all kind of visitors, as a critical mass of visitors is needed to change or strengthen the image of a city. If cities take a long-term view in this respect and do not hastily rule out any visitor category, cities and companies may indeed have converging interests concerning their target groups. However, if both actors tend to take a rather short term (income oriented) view, companies might be inclined to prefer local target groups (for example, to influence their consumer behaviour, or to attract potential employees), while the international visitor could be very attractive for the city economy. According to Williams and Shaw (1992) foreign visitors usually spend much more money per capita, when compared with domestic tourists. Of course, this has to do with the fact that domestic visitors usually don't need to pay for accommodation – for example because they live nearby or they spend the night at a friend's place. Foreign visitors usually spend more time in a place, which is also why they spend more money (as indicated, accommodation is crucial in this respect). However, this view is very much oriented towards short term income and hence misleading. Image effects must have a central position in the perception of cities and companies – a view that is oriented on short term profit only can undermine the understanding of converging interests of public and private in the realm of industrial tourism.

Nevertheless, income is, as we have seen, also a key goal of industrial tourism, and is a crucial determinant for the convergence of interests from public and private. By this we mean that for the city, positive image effects alone might not be enough: for a city it might be beneficial to attract some of the cash cow visitors (for example international visitors and overnight stayers) as well, in order to achieve a higher income. We assume that the situation is similar for companies, with the difference that the emphasis of companies will be more on long term benefits (unless a company operates an adjacent shop where its products or souvenirs can be purchased). The problem is that every visitor group delivers different kinds of benefits to cities and to enterprises. Hence, we assume that cities and companies need to engage in a negotiation process in order to agree on those target groups that bring most benefits in terms of income and image to both sides.

It may depend as well on the availability of investment to develop the necessary infrastructure (hotels, attractions, and so on) if a city economy can reap the benefits of (industrial) tourism to the full, and hence if a city's interest in industrial tourism is converging with the interests of companies. Industrial tourism as such is a great opportunity for cities, but cities need to take into account a wide array of additional conditions. Cities need to hold the ability to supply the needs of tourists, whether for food, souvenirs or hotel beds (based on Williams and Shaw, 1992). This means that cities cannot focus on the narrow field of industrial tourism alone, if they want to reap the full benefit in terms of income (and image).

To give an illustrative example of the integrated character of industrial tourism, we focus on the importance of hotel capacity here. As Williams and Shaw (1992) note, it is important to pay special attention to the accommodation sector, because it is accommodation – more than any other component – which has a major influence on the types of tourists attracted, the duration of their visits and the types and the quantities of jobs and income that will be created in the region (Williams and Shaw, 1992). If a city tries to develop a specific tourism niche, geared at specific target groups, it should make sure that there is sufficient hotel capacity in place, in a category that fits the needs of the target group. Students, for example, will rather show a preference for low cost accommodation, such as youth hostels or hotels of the lower categories. If these people are interested in the attractions a city has to offer, but cannot find accommodation that fits their needs, they will either choose not to visit the city, or they will opt to spend the night in a nearby place that does offer a sufficient supply of hotels in the required category. What does this mean? Well, on the one hand this means that cities need to be well aware of their free capacities in terms of hotels, and especially which category is suitable for which target group. It might be necessary for a city to stimulate the expansion of hotel capacities (of specific categories), always keeping in mind the preferences of the target group that is aimed for.

In this relation, one should be aware of the difference between excursionists and tourists (see Van der Borg, 1991). Tourists are defined as overnight stayers, whereas excursionists are visitors that visit a place, but spend the night elsewhere. If there are too many excursionists in a place and to little overnight stayers, this might be harmful for a city. Excursionists spend less money than overnight stayers, but also create costs for the city (for example congestion costs and the recycling cost of the waste they leave behind). Thus, cities need to reach a balance between these two groups, whereby a higher share of overnight stayers is more beneficial for them in terms of income. For a company it might not matter much if a visitor spends the night in the city or if the visitor comes as an excursionist; what seems to be important for companies is that the tourist pays a visit to their company and leaves with a positive impression. However, one could argue that spending the night in a city increases the probability that tourists visit more than one company, just because they have more time available in the place. Hence, companies and cities may have converging interests in this issue as well. The concept of excursionists and tourists does not mean that excursionists are per se bad; on the contrary, also this group can be very beneficial for cities. Nevertheless, cities have to aim at a healthy balance between excursionists and overnight stayers.

4 Our frame of analysis

In this book we define industrial tourism as visits to operational firms of people who are interested in the firm, its products, the production processes and/or the way a company is managed. We acknowledge that the term industrial tourism is a bit misleading because it not only involves visits to industrial firms, but to companies in other sectors as well. Also visits to publicly-owned companies are object of study, although we have to realise that their motives differ from the motives of privately-owned firms. Furthermore we emphasise that industrial tourism not only comprehends visits by (leisure) tourists, but by other customer groups as well. We can distinguish the following target groups:

- Leisure tourists. Industrial tourism could be particularly interesting for more experienced tourists who have seen many cities in the world, and who are now looking for truly unique experiences. Company visits provide an opportunity to see how a city is actually functioning, reflecting the identity (or personality) of a city.
- Students. For students, industrial tourism provides opportunities to learn more about production processes, while getting to know potential future employers.
- Professionals. This category includes suppliers, (business to business) customers, competitors, financiers and other stakeholders of a company. Suppliers, customers and financiers may want to learn more about the firm they cooperate with, while competitors and other firms are usually interested in innovations or the concept of industrial tourism as such. Moreover, also visitors of trade fairs and congresses belong to this group. The segment of 'incentive travels' also forms part of professional visitors.
- Journalists and researchers. Their motives are comparable with the motives of competing firms: they want to learn more about innovations, either in the core business or about the company visits themselves.

The aim of this research is to shed light on the following question: 'Under what circumstances can industrial tourism make a strategic contribution to the competitiveness of city and enterprise?' Based on this deliberation, we construct a frame of analysis in this section, which will be applied to the case studies. We need such a frame to analyse different forms of industrial tourism in a systematic way. With the help of the literature review above, we could identify four important issues that need to be taken into account for securing the support of both city and enterprise, and ultimately for making industrial tourism work in practice: (1) adequate target group selection; (2) attractiveness of the industrial tourism product; (3) benefits in terms of income and image for city and enterprise; and (4) the organisational form.

Adequate target group selection

The scope and scale of the target group are strongly related to the attractiveness (and accessibility) of the product. The scope concerns the characteristics of the target group: notably their age (for example children, young adults or older people), their travel motives (for example leisure, business or education), and also their educational level, income level, sex, civil state, and so on could be relevant. In this relation also the distinction between excursionists and tourists (Van der Borg, 1991) is crucial. With the scale of the target group we refer to the geographical size of the target group, for instance making a distinction between local, national and international visitor attractions. Some industrial tourism products might be interesting for a very small niche market only, while other products have the potential to attract mass tourism. We already elaborated on this first condition in section three. The attractiveness of an industrial tourism product is a subjective notion and strongly depends on the preferences of the target group.

Attractiveness of the industrial tourism product

Taking a look at the theory we expect that for some firms it is easier to attract visitors than for other firms, simply because their products (or processes) are more appealing. Examples of such products are cars, motorbikes, chocolate, wine, and movies. Product characteristics are particularly relevant for leisure-oriented visitors who want to taste and/or touch consumer products. What also matters in our view is the attractiveness of a production process. An example of a factor that plays a role in this respect is the degree of innovation: highly innovative production processes might attract another (though not necessarily smaller) target group than more traditional methods of production. The characteristics of the production process have a strong influence on the demand from students, professionals, journalists, and researchers.

Besides the quality of the product, also the quality of the visitor services should be taken into account. The demand and attractiveness for an industrial tourism product can be enhanced by providing special visitor facilities, such as (interactive) information panels, (professional) guides, opportunities to feel, taste or touch the product (or inputs), presentation facilities, catering facilities, and so on. The required visitor facilities and their quality level strongly depend on the target groups to be attracted.

The issue of accessibility is also highly important for the attractiveness of a product. On the one hand this refers to the ease of reaching a destination, on the other hand accessibility can also refer to the availability of other primary/secondary tourism products nearby. The attractiveness of a good/company might be increased if other attractions are located nearby, which makes it more attractive to travel to a location that is not so accessible. This could mean that rather remote companies should seek ways to team up with nearby tourism attractions, as visitors can combine several experiences in one trip. This calls for an integrated approach of tourism. For cities, the challenge is to create packages that reflect their image and/or identity, and that respond to the needs of customer groups. Packages have to be attractive and accessible, just like the individual products.

Also the accessibility on the production site is a relevant factor, with the disabled as a target group that deserves special attention. Accessibility can however also be interpreted in terms of legibility – the ease with which visitors are able to find a location – which requires adequate communication and information provision (in the city and on the internet). In this respect signposting plays an important role: standard road signage can be beneficial.

Benefits in terms of income and image for city an enterprise

To reach converging interests of the public and private side, one needs to ensure that the benefits overweigh the costs for both parties. With benefits we refer to increases in income and in image, costs refer to the initial investments that have to be made, and potential risks that might occur. Industrial tourism can make a strategic contribution to the competitiveness of city and enterprise if the product responds to the demand of particular target groups, and if city and companies can agree on a number of common target groups.

As a first condition, a tourism product is needed that benefits the city in terms of income and image. In this respect several aspects have to be taken into account. Highly populated areas could be very fertile places for industrial tourism, because they offer a sufficiently big critical mass of companies that can ensure the supply of industrial tourism. Industrial tourism is a rather new concept, and thus not so many regions have adopted an industrial tourism strategy yet; such a strategy may help a region to reach a competitive advantage over other regions. Cities may exploit the benefits to the full if they hold the ability to supply the needs of tourists, whether for food, souvenirs or hotel beds. Especially accommodation is crucial in this context because it has a major influence on the type of tourists you attract and on the length of their stay. In this respect, cities could try to increase the share of tourists that spend the night in their city, as this category of visitors generally spend more money during their stay. Especially international visitors are an attractive target group for cities, as they spend – on average – more money than domestic visitors, thus generating more benefits for the city. However, also local tourists and excursionists are beneficial for cities, as these groups can help to turn around the image of the city. Improving the image of a city is a big task, and the more people you can reach, the quicker it can be accomplished.

As we already mentioned in section four, companies can achieve several benefits by being engaged in industrial tourism. Industrial tourism can be used as a marketing tool for their products (both for consumer products and intermediate products) and may present a way to be closer to the customer's needs. Opening the company's doors to visitors is also a way to show that corporate social responsibility is taken seriously. Moreover, company tours can help to build pride among employees and this way support the daily production process. It can also generate income (from the sale of tickets and souvenirs) and it can help to attract employees.

Organisational setup

During the literature review it became clear to us, that most recommendations concerning organisational aspects concentrate on companies. 'How can companies organise industrial tourism?' is the main question that is dealt with in this respect in the literature. This is mainly about organising industrial tourism in a way to overcome encountered barriers, and to avoid negative effects that could occur due to company visits. The organisational framework needs to ensure that a product meets safety, security and other basic requirements. If the actors are aware of these issues, it is usually very easy to find innovative ways to tackle them. Every company will face a certain capacity limit: a maximum number of visitors that can tour a company during a defined timeframe under the present conditions. In many cases these limits can be increased by setting relatively easy measures. However, very often it is rather the unwillingness of the company – and not these barriers – that restricts the number of visitors. Moreover, companies need to be aware of the seasonality of tourism and the problems that might come along with this if not properly managed – again, these problems can be solved by employing some anticipatory thinking. We acknowledge the importance of these kind of organisational aspects in companies, but our view on industrial tourism is somewhat broader in this respect, and includes organisational issues in the city and the major actors in the city as well. We suggest that industrial tourism can be strengthened, if a city is able to create a conducive organisational environment for companies.

To conclude

The literature on industrial tourism suggests that there has been a growing demand in industrial tourism over the past decades, and predictions proceed on the assumption that this development will continue. More and more companies open their doors for visitors, and with a certain time lag, also cities have realised the benefits they can reap from well organised company visits. The main drivers for cities and companies are improvements in image and income. In this sense, we can observe converging interests between companies and cities. For companies, a city with a strong economic image is very positive; and also for the cities it is highly beneficial if they harbour an array of companies that make use of industrial tourism to boost their image and income. Industrial tourism can be a true win-win situation: if it is organised in a professional way, if the quality and attractiveness of the company tours is assured, and if cities and companies can agree on those target groups that bring most benefits to both sides.

In the next chapters we will apply this research framework to the case studies. Each case study has a similar structure: after an introduction (the first section), the second section addresses the general profile and the tourism profile of the examined city and its region. In section three we analyse the organisational framework for industrial tourism in the selected city/region, followed by an overview of industrial tourism products in section four. These observations will be confronted with the frame of analysis (the four conditions that we described above) in Section five.

References

Archer, B. 1977. *Tourism Multipliers: The State of the Art.* Cardiff: University of Wales Press.

Baumann, B. 1999. *Bestandsanalyse des Industrietourismus zu produzierenden Unternehmen in Suedwesten Deutschlands*. Trier: Europeaeisches Tourismus Institut Gmbh.

Berg, L. van den and Braun, E. 1999. Urban competitiveness, marketing and the need for organising capacity. *Urban Studies*, Vol. 36 (no. 5–6), pp. 987–999.

Brumberg, B. and Axelrod, K. 1995. *Watch It Made in the USA: A Visitor's Guide to the Companies That Make Your Favorite Products*, 1st ed., Sante Fe, NM: John Muir Publications.

Carter, J. 1991. Watching work go by. *Environmental Interpretations*, Vol. 6 (no. 4), pp. 10–11.

Frew, E.A. 2000. *Industrial Tourism: a conceptual and empirical analysis.* PhD Thesis, Victoria University.

Gelbert, D. 1994. *Company Museums, Industry Museums and Industrial Tours: A Guidebook of Sites in the United States that are Open to the Public.* Jefferson: McFarland and Company.

Harris, F. 1989. From the industrial revolution to the heritage industry. *Geographical Magazine*, Vol. 61, pp. 38–42.

Henly, A. 1988. Industry shows its face. *The Engineer*, 27 October, p. 28.

Hill, R.M., Alexander, R.S. and Cross, J.S. 1975. *Industrial Marketing.* Homewood IL: RDIrwin.

Kalinowski, K.M. and Weiler, B. 1992. *Special Interest Travel.* London: Belhaven Press.

Li, L. and Soyez, D. 2006. Industrial tourism destination management in Germany: A critical appraisal of representation practices, in: *Community Tourism and Border Tourism*, edited by BAO Jigang, XU Honggang and Alan Lew. Beijing: China Travel Publisher, pp. 408–29.

Maccannel, D. 1976. *The Tourist: A New Theory of the Leisure Class.* New York: Schoken Books.

Mader, T. 2003. *Produzierende Betriebe als touristische Attraktionen im Ruhrgebiet. Grundlagen, Erscheinungsformen, Probleme. Magisterarbeit, Heinrich-Heine-Universitaet Duesseldorf.* Hamburg:Diplomica Gmbh.

Marcon, A., Preuilh, P. and Ksouri, S. 2000. Tourisme de découverte économique et visites d'entreprises, in: Conseil national du tourisme, ed. *La Documentation française*, pp. 10–16.

Martin, B. and Mason, S. 1993. The future of attractions: Meeting the needs of the new consumers. *Tourism Management*, Vol. 14 (no.1), pp. 34–40.

Mitchell, M.A. and Orwig, R.A. 2002. Consumer experience tourism and brand bonding. *Journal of Product and Brand Management*, Vol. 11 (no. 1), pp. 30–41.

Murphy, P.E. 1982. Cited from: Williams, A.M. and Shaw, G. 1992. *Tourism and Development: A European Perspective.*

Prentice, R. 1993. *Tourism and Heritage Attractions*. New York: Routledge.

Schmidt, D. 1988. Industrie-Tourismus: Moeglichkeiten und Grenzen einer Anwendung auf ausgewaehlte Gemeinden der Landkreise Coburg, Kronach und Lichtenfels, Bayreuth. *Arbeitsmaterialen zur Raumordnung und Raumplanung*, no. 63.

Soyez, D. 1993. Kulturtourismus in Industrielandschaften, in *Kulturtourismus in Europa: Wachstum ohne Grenzen*, edited by Becker, C. and Steinecke, A., ETI-Studien, 2, Trier, pp. 40–63.

Swarbrooke, J. 2002. *The Development and Management of Visitor Attractions.* 2nd ed. Burlington: Butterworth-Heinemann.

Van den Berg, L., Braun, E. and Otgaar, A.H. J. 2003. *City and Enterprise; Corporate Community Involvement in European and US Cities*. Aldershot: Ashgate.

Van der Borg, J. 1991. *Tourism and Urban Development.* Amsterdam: Thesis Publishers.

Van der Borg, J. 1994. Demand for city tourism in Europe: tour operater's catalogues. *Tourism Management*, Vol.15 (no. 1), pp. 66–69.

Van der Borg, J. 1995. *Tourism and Cities of Art UNESCO/ROSTE Technical Report No. 20*. Venice: UNESCO.

Williams, A.M. and Shaw, G. 1992. *Tourism and Development: A European Perspective.* London: Belhaven.

Yale, P. 1991. *From Tourist Attractions to Heritage Tourism.* Huntingdon and/or Yale, P. 1992. *Tourism in the UK.* Huntingdon.

Chapter 2
Autostadt, Wolfsburg

1 Introduction

In this case study we provide a description and analysis of industrial tourism in the German city of Wolfsburg, with a particular focus on the most dominant local actor in the realm: the Volkswagen Group, which has its global headquarters in Wolfsburg. Volkswagen's industrial tourism activities are commonly seen as best practices, and various authors – among them Mader (2003) – acknowledge them as such. The Wolfsburg case is particularly interesting because it already is in a very mature stage of development. Moreover, cars are an extremely attractive consumer product, which is why the interest in car manufacturers has been high since the early days. Car manufacturers were 'naturally driven' to develop company tours and this is probably also the reason why the car industry today is a forerunner when it comes to the quality of company visits. The industry has a long history and ample experience with it. In Wolfsburg, traditional company visits have been complemented by the theme park-like Autostadt ('Car-city') in 2000, which is a subsidiary of the Volkswagen Group.

As the present case study has a strong focus on one company, its structure is slightly different in comparison with the other cases. After this introduction, section two deals with the characteristics of Wolfsburg, followed by an introduction about Volkswagen in section three (instead of the organisational framework of industrial tourism). Section four comprehends a description of the industrial tourism products in the city of Wolfsburg that are operated by the Volkswagen Group: the factory tours in the Volkswagen car plant, and the Autostadt theme park: a kind of brand land that represents the sub brands of the Volkswagen Group. In section five we confront the Volkswagen case with our research framework, drawing some conclusions on the conditions under which industrial tourism can be developed.

2 City profile

History, location and demography

Wolfsburg is a very young city and has its origins in the Second World War; the city was basically planned by the National Socialist German Workers' Party (NSDAP) as a town for the workers of the Volkswagen factory, and was built from scratch in 1938 in the open countryside. The location was chosen in view of strategic considerations: the site was located right in the centre of the German Reich and had good communications to the railroad system and to waterways (City of Wolfsburg, 2001). The artificial waterway *Mittellandkanal* (321 km long) passes through the city and links the Ruhr area with Berlin, and connects many important industrial areas along the waterside. This waterway is also an important determinant for the internal accessibility of Wolfsburg, as it cuts the city in two: the city centre is located south of the channel, while the Volkswagen premises are located to the north of the waterway. This strict separation of functions was an integral feature of the original plans for the new town Wolfsburg (City of Wolfsburg, 2001).

In its very beginnings, the name of what is today Wolfsburg was actually *Stadt des KdF Wagens*, which means literally City of the KdF Car. The KdF car was one of Volkswagen's most successful products, later to be known as Volkswagen Beetle. KdF stands for *Kraft durch Freude*, which can be best translated as strength through joy. In 1945 the city received its actual name. The name was derived from Wolfsburg castle, which is located within the municipal boundaries of the city. Wolfsburg is located a mere 90 km east from Hannover (the capital of the state Lower Saxony), and around 230 km west from the German capital Berlin. Since 1998 the city is also connected by high speed train (ICE) to both of these important centres, whereby the trip to Berlin takes only one hour. The city is (obviously) also connected to the German highway system, by the A39, an 'offshoot' of the main highway A2, which connects Hannover and Berlin. Both Berlin and Hannover have an international airport. A smaller regional airport is located at 30 km distance, near the city of Braunschweig.

After 1945 Wolfsburg experienced a sharp growth: while in 1945 the city had 18,677 inhabitants, this number rose to 93,494 in 1970. In 1972 some neighbouring towns were merged with Wolfsburg, resulting in a steep increase of the population, to 130,979. In 1980 Wolfsburg reached its population peak with 131,225 inhabitants; by 2008 Wolfsburg's population had shrunk to 120,078 inhabitants, of which 11,787 (9.8 per cent) were foreigners (City of Wolfsburg 2008a). A high share of 5,300 of these foreigners are Italian citizens (City of Wolfsburg 2008b). Volkswagen has a long tradition of attracting Italians (but also other foreigners) as workforce for their factories.

Economic structure

In 2008 the unemployment rate in Wolfsburg amounted to 7.8 per cent (City of Wolfsburg, 2008). The overall number of jobs in Wolfsburg was 91,714 in the year 2006; of these, 61,807 jobs were provided by the production sector, 9,989 by commerce, catering and traffic, and another 19,662 by the service sector. Agriculture, forestry and fishing together employed only 255 people (City of Wolfsburg, 2006). Apart from Volkswagen's facilities, there are actually not many companies in the production sector in Wolfsburg. The most important industrial companies – apart from VW – are subcontractors of Volkswagen. Most of these subcontractors are located in industrial estates that were designated especially to them. Today Wolfsburg is – after the city of Hannover – the second biggest centre of employment in the German federal sate of Lower Saxony (City of Wolfsburg 2008a). Its strong economic position is also shown by the fact that 55,957 people commuted to Wolfsburg in 2006, whereas only 5,807 persons commuted from Wolfsburg to their place of work located elsewhere (City of Wolfsburg, 2006).

The omnipresence of Volkswagen in Wolfsburg also means that Wolfsburg's economy is dominated by the company. The city and Volkswagen are fully aware that their fate is tied to a big extent, which we can also see in common efforts to overcome the crisis that hit Volkswagen in the early 1990s. In their common strategy – the so called 'Autovision' – and through a common organisation that was founded in 1999 – the Wolfsburg AG – the city of Wolfsburg and the Volkswagen AG commit themselves to raise job security in Wolfsburg. The Wolfsburg AG is a public private partnership between Wolfsburg and VW, whereby every partner invests 50 per cent of the capital. The goal of this strategy and the resulting actions is to increase the importance of service and leisure industries in the city, and to attract private capital. Additionally to these efforts Volkswagen opened the theme park Autostadt in the year 2000. These initiatives were accompanied by a decline of unemployment (City of Wolfsburg, 2001).

Wolfsburg is also a student city: it hosts a branch of the Braunschweig/Wolfenbuettel University of Applied Sciences which offers a wide range of economic and engineering studies (among others vehicle construction), but – for example – also a study path about health care. Apart from this institution, Wolfsburg also hosts the AutoUni, which is an institute for continuing education within the Volkswagen Group, entitled to award master degrees. This institute is mainly geared at senior specialist staff and managers of the VW group worldwide. There are plans to admit also interested students from VW's suppliers and partners, and by 2010 the university will be open to all interested parties.

Tourism profile

From a traditional tourist point of view, Wolfsburg cannot be conceived as a very attractive city on the first sight. The city lacks historic buildings, and less than 5 per cent of all visitors come to Wolfsburg because of its cultural amenities and art museums. However, Wolfsburg boasts a number of different and rather new attractions:

- The Autostadt theme park: opened in 2000, this theme park is the biggest tourist draw in Wolfsburg.
- The Volkswagen car museum: since 1985 this museum displays 140 historic Volkswagen models, among them the famous VW Beetle from the Herbie movie series.
- The Volkswagen car plant: since 1948 this production facility is open for visitors. Wolfsburg's main landmark, the Volkswagen power plant with its four characteristic brick chimneys, is a characteristic sign for the nearby Volkswagen factory.

- The Planetarium Wolfsburg: this installation was opened in 1983 and is one of the ten biggest planetariums in Germany. It is owned by the city of Wolfsburg.
- The Kunstmuseum Wolfsburg: this private art museum opened its doors in 1994 and is specialised in modern and contemporary art.
- The Volkswagen Arena: this stadium was built in 2002 and is the homestead of the soccer team VFL Wolfsburg, which plays in the first German Bundesliga.
- The Badeland and the Allersee: Badeland is a big swimming and sauna complex, whereas the Allersee is a lake for bathing with facilities for waterskiing and wakeboarding.
- The Phaeno Science Centre: this experimental museum was designed by star architect Zaha Hadid and opened its doors in 2005.
- The Designer outlet centre: this complex of stores is located very close to the inner city and is the most recent addition to Wolfsburg's attractions (inauguration in 2007).

As we can see, three of Wolfsburg's main attractions (Volkswagen Plant, Car Museum and Autostadt) are related to industrial tourism. In addition, there is one other large company in Wolfsburg that offers company tours on a regular basis: Thyssen Fuegetechnik. Moreover, Wolfsburg hosts an exhibition of subcontractors of the automotive industry every two years, which is within Germany the second biggest only to the automotive exhibition in Frankfurt. This fair attracts a lot of visitors from Asia, particularly from China. Due to the high demand for company visits, especially from Asian fair visitors, also some of VW's subcontractors open their doors for visitors during this period. While the Volkswagen plant and the Autostadt attract visitors from all over Germany and the world, most of the other attractions rather draw visitors from the region. However, also the Phaeno Science Centre (which attracted around 700,000 visitors in the year of inauguration), the art museum and the Volkswagen Arena are drawing visitors from all over the nation. Apart from the Volkswagen factory, all the other attractions are rather recent additions to Wolfsburg's tourist products (within the last 25 years). Before the year 2000 the only big attraction in Wolfsburg was a tour in the Volkswagen plant itself, and several guidebooks even recommended tourists to skip Wolfsburg altogether on their travel itinerary.

It is worth noting that before the year 1989 tourism in Wolfsburg was hampered by the fact that the city was located at the periphery of the Federal Republic of Germany. Wolfsburg's location, only 6 km from the Iron Curtain, was indeed not very favourable. Wolfsburg's natural catchment area for regional tourists was literally truncated. With the collapse of communist Germany in 1989, and the reunification of Germany one year later, also Wolfsburg's tourism slipped into gear. The excellent accessibility to Berlin has probably spurred the number of visitors even more: Wolfsburg has become a popular stop of tourists heading to the federal capital of Germany. The increased accessibility has, however, not only brought blessings for Wolfsburg: as Berlin is only one hour away, a visit to Wolfsburg can be easily done as a daytrip nowadays, without any need to spend the night in the city. Already before the year 2000 leisure tourists were coming to Wolfsburg mostly as excursionists (visitors that do not spend the night in the city and hence leave less money in the city), but with the advent of the high speed train, the share of visitors that spend the night in Wolfsburg has decreased. A (relative) decrease could also be experienced in the segment of business travellers. Business travellers have always been a very strong segment of Wolfsburg's visitors and not surprisingly most of them come because of Volkswagen.

In 2006, the city of Wolfsburg boasted an overall number of 2,764 hotel beds (Niedersächisches Landesamt für Statistik, 2006). The annual statistics on tourism recorded 377,993 overnight stays for the year 2006. Of these, 300,000 can be attributed to German citizens, which clearly indicates the overwhelming importance of the national visitor market. Nevertheless, 42,477 bed nights can be attributed to other European countries (most notably Spain, Portugal, the UK, the Netherlands, Poland, Austria and Italy), 10,715 to Asia (around half of these to China) and another 12,193 to the Americas (City of Wolfsburg, 2006). There have been several marked increases in overnight stays in the last years: in the years 1999 and 2000 an increase of overnight stays could be observed, due to the arrival of staff of the newly established Autostadt. Also in 2003 Volkswagen was responsible for a rise in bed nights; that year, VW hired a lot of external workers for a temporary project, some of which were accommodated in Wolfsburg's hotels (453,522 overnight stays in 2003). Clearly, also the product presentation of a new VW Golf model had a positive impact on the 2003 statistics. In the following years, overnight stays returned to a somewhat lower level.

A general problem of Wolfsburg is the bad hotel occupancy rate in weekends: a typical feature of cities with a lot of business travel. Wolfsburg's hotels and Wolfsburg Marketing are trying to oppose this by attracting package tourists

Table 2.1 Overnight stays in Wolfsburg, 1999–2006

Year	Overnight Stays
1999	259,337
2000	326,271
2001	304,739
2002	372,290
2003	453,522
2004	402,959
2005	339,960
2006	377,993

Source: Wolfsburg Marketing, 2007b

especially at the weekends. Wolfsburg Marketing itself also acts as a travel organiser; they sell trips to Wolfsburg in collaboration with travel agencies (most notably coach tour agencies and organisations specialised in group trips) in Germany and abroad. Around 90 per cent of these package trips to Wolfsburg involve a visit to the Autostadt theme park. The promotional activities of Wolfsburg Marketing are taken throughout Germany, with a focus on the nearby large cities Berlin and Hamburg. Apart from acting as a travel and promotion agency, the Wolfsburg Marketing organisation sees its main tasks in the cross linking of actors in the realm of tourism, and in facilitating the combination of various attractions – the aim of this strategy is to tap into target markets that the single attraction cannot access on its own. This way, customers can assemble their own visit programme, by combining some of the modules offered by Wolfsburg Marketing. Volkswagen also expressed a clear interest in establishing more partnerships with other attractions in and around Wolfsburg and hereby supports the city's aims. The main aims of these actions are to attract as many people as possible to Wolfsburg and to increase the number of overnight stays. Another aim of the Wolfsburg Marketing organisation is to make German people realise that Wolfsburg is more than just Volkswagen, highlighting hereby other attractions that are not related to industrial tourism. On the other hand Wolfsburg Marketing clearly relies on the prominence of Volkswagen on the international tourist markets.

3 Volkswagen

As we have already discussed the economy of Wolfsburg is still very much dominated by the Volkswagen Group; but Volkswagen does not only influence the economy of Wolfsburg, it is an omnipresent element in many other areas as well. Wolfsburg's promotion efforts demonstrate for example the strong ties to Volkswagen: in 2003, the city renamed itself to Golfsburg for one week, due to the market launch of a new Volkswagen Golf model. Returning to the very beginnings of Volkswagen we can state that, just like the city of Wolfsburg, also the Volkswagen group has its origins in the Second World War, being founded in 1938. The history of the city and the company are deeply interwoven. During the war the Volkswagen factories produced jeeps, aeroplanes and military equipment; the workforce in late 1944 was made up of around 18,000 people, two thirds of whom were forced labourers. By 1945 only around 6,000 employees were still employed at Volkswagen's Wolfsburg location (Volkswagen, 2007). What followed after 1945 was a remarkable expansion of Volkswagen – both in Wolfsburg and on the international level.

Today, Volkswagen alone employs around 43,500 people in Wolfsburg; of these, a major part of 9,000 employees work in research and development. Moreover, a few thousand people from temporary employment agencies also work at the plant; so Volkswagen is actually responsible for an employment of more than 50,000 people, which accounts for around 55 per cent of all jobs in

Wolfsburg. On a worldwide scale the Volkswagen group operates 44 branches (of which 13 are located in Germany) and employed 328,594 people in 2007, of which 175,206 in Germany (Volkswagen, 2008). With 43,500 jobs in Wolfsburg – more than 13 per cent of the worldwide VW staff – Volkswagen Wolfsburg is also a heavyweight on a worldwide scale. The importance of VW Wolfsburg is not only expressed in the amount of personnel: also the global headquarters of the Volkswagen Group are located in Wolfsburg. Volkswagen is Europe's largest car manufacturer, whose market share amounted to 19.5 per cent in Western Europe. The Volkswagen Group held a 9.8 per cent share of the world passenger car market in the year 2007, corresponding to 6,192,000 sold cars. This number does not only contain the Volkswagen cars themselves, but also six further sub-brands of the Volkswagen Group: Audi, Bentley, Bugatti, Lamborghini, Seat and Škoda. On overall Volkswagen boasts eight sub-brands, including Volkswagen Commercial Vehicles (Volkswagen AG, 2008).

In the Wolfsburg plant, the car models Golf, Golf Plus, Touran and Tiguan are produced (Volkswagen AG, 2007); also the most important press shop of the Volkswagen group is located in Wolfsburg. The Wolfsburg plant is one of the most modern automobile plants in the world, which is for example demonstrated by the fact that welding is only done by robots and not anymore by people. In total, 1,140 robots (Merian, 2007) are active at Volkswagen. Every 20 seconds a new car is finished off in the Wolfsburg plant, which makes 3,000 cars per day, assuming the plant is operative on a three shift basis. Oddly enough, even more cars were produced per day in former times: around 5,000 to 6,000 models of the old VW Beetle a day. One reason for this is that today cars consists of many more single components than 30 years ago; another reason is that the standard car does not exist anymore: there are 300,000 possible variants of the VW Golf, for example.

4 Industrial tourism products

Volkswagen Wolfsburg operates two industrial tourism products: the factory visits of the Volkswagen plant and the Autostadt Theme park. In organisational terms these two products are not integrated. The factory tours are organised by Volkswagen's department of Visitor Services, which reports directly to Volkswagen Communications Brand and Product (part of the Volkswagen AG stock corporation). Autostadt is a subsidiary of the Volkswagen Group, and is intended to represent all sub-brands of the group equally.

Visiting the Volkswagen plant

The Volkswagen plant in Wolfsburg has a long history in company visits. Already in 1948 the carmaker decided to open its operating facilities to the public on a regular basis. Since then, more than 6 million visitors have witnessed the production process of Volkswagen. Since the year 2000, more than 1.7 million people have participated in a factory tour (Volkswagen Visitor Services 2007b). Table 2.2 shows that visitor numbers have been climbing constantly from 1948 till around 1970; that year already around 120,000 visitors per year had a visit to the Volkswagen factory. Afterwards the visitor numbers decreased considerably and it was only in the year 2000 – in the course of the opening of the Autostadt theme park and the resulting increased attention for Volkswagen – that visitor numbers could cut the record of the late 1960s and 1970s. Since the year 2001, visitor numbers have been constantly above 180,000 visitors per year. The year 2001 marked a peak in Volkswagen's history, with 262,119 annual visitors. On top of that, Volkswagen potentially could have had even more visitors in that year. The rush was so high that VW had reached its upper limit and even had to reject people. As we can see below, Volkswagen classifies visitors in leisure visitors (that is, visitors that are channelled into the plant via the Autostadt) and special visitors. In the special visitor group, VIPs, delegations from importers, dealers and universities (most notably student groups) clearly account for most visits. Some German universities even incorporated the visit to the Volkswagen plant into their

Table 2.2 Visits to the Volkswagen Plant Wolfsburg, 1948–2006

Year	Leisure visitors and people picking up their cars	Special visitors (VIPs, journalists etc.)	Total number of visitors
1948	/	/	25,628
1950	/	/	36,419
1955	/	/	83,684
1960	/	/	99,283
1965	/	/	115,623
1970	/	/	121,623
1975	/	/	77,926
1980	/	/	80,151
1985	/	/	92,942
1990	/	/	104,869
1995	/	/	57,871
1999	/	/	40,945
2000	118,993	8,988	127,981
2001	238,720	23,399	262,119
2002	225,192	23,958	249,150
2003	228,102	24,914	253,016
2004	215,527	23,341	238,868
2005	213,912	23,760	237,672
2006	163,989	22,951	186,940
2007	158,269	26,968	185,237

Source: VW Visitor Services, 2007b

teaching curricula. By 2007, 38 factory tours were conducted on average per day, whereby Volkswagen admits a maximum of 1,300 people to its premises per day during summertime. Of these 1,300, usually 1,000 are leisure visitors and people that pick up their cars (all of which are also Autostadt visitors), and around 300 are special visitors; in the winter months tours for a maximum of 1,000 guests per day are offered (Volkswagen Visitor Services 2007b).

Learning from competitors Before the year 2000 – when Volkswagen upgraded the quality of the factory tours in the Wolfsburg plant – the people in charge had been visiting their competitor's facilities (Porsche, Audi, Opel, Ford, BMW & Mercedes-Benz) throughout Germany in order to observe their best practices and to avoid making the same mistakes. One of VW's competitors had for example acquired a number of panoramic trains for showing around visitors; but instead of providing live comments by a guide and giving a possibility to interact, this company only used information screens in the trains. Obviously, there is not much personal touch in this, and VW detected that visitors might prefer to get information from a real person that actually works in the plant in question. Volkswagen also detected another mistake in its competitor's tour: visitors were shown around when the machines were not running, and seeing an empty workshop is considered not the most exciting experience. The danger that lies in providing tours with low quality is that people might become angry about the tour that is offered to them. So if not done properly, the whole idea of company visits can actually backfire and damage the image of a company. The outcome of the pre-2000 competitor screening had actually confirmed Volkswagen's willingness to invest and its emphasis on quality. Volkswagen is still trying to learn from its competitors; they initiated a platform which brings together representatives from Volkswagen, Ford, BMW, Opel, Daimler and others on a regular basis (once or twice a year) for an exchange of expertise and for participating in workshops about visitor management.

Professional guides The constant stream of visitors requires professionally managed visitor services and professional guides. It is Volkswagen's policy that only VW employees are employed as guides for factory tours; no external guides are allowed to guide a group around. Although the commitment to full employment of the guides means more salary costs for VW than an array of flexible external guides that are available on request, this policy ensures that the guides are knowledgeable about the production process, and other topics touched upon during the tour. The guides have the freedom to gain knowledge in the way they like and VW encourages them to steer their own further education. For this reason and for the sake of improving the company tours, the guides get 'free time' to speak to people in other departments and to identify areas that could be interesting for visitors, and to learn what they are allowed to communicate to the visitors and what not. Most of the times the initiative to introduce new special tours comes from the guides, who are also responsible for the development and preparation of these tours. It is not a coincidence that the guides themselves are responsible for improving and introducing company visits: they are the ones who are closest to the customers, and hence they can be also expected to know what kind of experiences the visitors request. Volkswagen's tour-guides are professionals that are trained for this job. The guides work on a full time basis (they do not have any other 'regular job' in the company, like it happens often in companies that have not professionalised their visitor services). However, most of the tour guides used to work at the assembly line of the plant, and were later on recruited internally by the department of visitor services. The fact that most of the guides have formerly worked in production for at least one year prepares the ground to ensure that a realistic and vivid picture of the production processes can be conveyed to the visitors. Even the head of the department of visitor services himself has worked for several months on the assembly line at the beginnings of his career at Volkswagen, which he considers a valuable experience, mostly because it is important in this field 'to understand the basis'.

No standard tours Volkswagen's philosophy is rather down-to-earth and tries to avoid an artificial show for the visitors. Moreover, the aim of Volkswagen is not to provide a standard tour for every visitor, but to bring in some variation. This is also reflected in the training of the professional guides: most guides are trained to be specialists in one specific area of the factory. Even though, each guide has to conduct an array of all different kind of tours, in order to prevent that the work becomes monotonous. Volkswagen tries to focus its company tours on the exciting moments of the production process (for example the press shop, assembly, test rigs and body construction). The visitor trains stop for a while at each highlight, so that visitors get the possibility to have a good look and to ask questions to the guide. The press shop can be seen as one of the attractions of the plant; it is one of the few press shops worldwide with such huge dimensions. Here are components (like side panels, floors and rear flaps) not only produced for the Wolfsburg plant, but also for the VW branches in Mexico, China and South Africa and for Audi, Seat and Škoda. These exciting moments of production are, however, scattered all over the plant which covers 6.5 km^2. As one can imagine, also the dimensions of some buildings are stunning: the longest building on the premises stretches over 2.2 km. In order to cover most of the highlights of the factory, a company visit usually takes the visitors on a 7 to 8 km long itinerary through the plant; this clearly indicates why guided tours on foot are out of question for Volkswagen.

Furthermore, the guided tours also provide information about the social conditions of the workers, and Volkswagen aims to give a human touch to the tours and to show the man behind the worker. Nevertheless, visitors do not have the possibility to talk to workers. Blue collar workers often lack the skills that are necessary for professionally communicating with visitors and unprofessional communication could have adverse implications for the image of any company. However, Volkswagen has plans to qualify certain workers to directly communicate with visitors during the factory tours.

Clearly, there is an emphasis on quality in Volkswagen's efforts to showcase its main factory. Twelve panorama trains – capable of carrying 30 guests – take around the leisure visitors and those customers that come to pick up their cars in the factory. Each of these trains costs around €600,000. Volkswagen has chosen deliberately for a relatively small group size of maximum 30 people, in order to ensure that information and Volkswagen's messages are conveyed properly and personally to the visitors. During production time, two panorama trains drive off every 15 minutes from the starting point of the tour for leisure visitors at the Autostadt theme park (Volkswagen Visitor Services 2007b). The vehicles are also allowed to use public roads, in order to bridge the 3 km distance between the theme park and the Volkswagen plant. These trains are used to serve the mass of visitors; nevertheless, these panorama trains do not follow a kind of standard tour through the Volkswagen plant. In reality, the panorama trains cannot take the same route every time, in order not to disturb logistics and the production process. Also the schedule and itinerary of the trains need to be synchronised with the logistics departments and the different branches of production. The duration of a normal tour is one hour, the tours for the clients that pick up their cars lasts one hour and 30 minutes.

Special visitors, special tours So called special visitors – VIPs, guests of the board of management, journalists, politicians, visitors from associated companies, importers, dealers, groups from universities and so on – receive an even higher amount of attention: Volkswagen has acquired four special, convertible-like Volkswagen Golf vehicles plus one VIP Touran. In order to ensure that the visitors can hear the explanations of the guide in spite of the occasional noise in the factory, a loudspeaker is placed in every head restraint. Unlike the leisure visitors, these special visitors do not take part in a standard tour but are offered a tailor-made tour around the Volkswagen premises. Volkswagen also uses this special treatment to deal with customers that have communicated their anger to Volkswagen, due to the troubles they had with their car (like malfunctioning). In order to please angry customers, Volkswagen invites them for a lunch and for a VIP tour of their plant. This way, Volkswagen wants to make sure that these people don't communicate their anger to other people; word of mouth can be very powerful, both in a positive and in a negative sense.

In the course of their special visitor services, Volkswagen sometimes also combines a visit to one of their subcontractors with their own plant tour, but only if visitors ask explicitly for it. Sometimes Volkswagen also gets requests from subcontractors, whose visiting delegations also want to tour the Volkswagen plant. As we have already indicated, Wolfsburg does not host a lot of companies in the production sector, apart from Volkswagen itself. Most of the subcontractors of Volkswagen are rather small companies, which do not show any particular interest in offering company tours, at least not on a regular basis.

Volkswagen offers a number of special factory tours throughout the year, among them a Logistic Factory Tour, a tour for all the new citizens of the city of Wolfsburg, a 'Green Factory Tour' and a special tour called 'Cars & Candlelight'. The 'Green Factory Tour' has been designed as a response to the increasing environmental consciousness of the public. Before this initiative, Volkswagen was actually already investing a lot in environmentally-friendly techniques, while this was not always evident enough to all stakeholders. This gave rise to the introduction of a special factory tour dealing with environmental issues, which started in summer 2008. This 'sustainability tour' will also bring visitors in touch with issues like water treatment and green techniques and will inform about Volkswagen's initiatives to restrict pollution. Another tour, so called 'Cars & Candlelight' is a 'VIP-tour' that has been developed by VW as an entertainment for special guests. This tour takes place in the evening, and is followed by an exclusive candlelight dinner in the Volkswagen press forum. This is the only tour for which VW actually charges money.

Apart from the regular special tours, Volkswagen also introduced a number of special off season tours that are deliberately organised in the winter period (January to April), when there are less regular visitors. The main reason for these tours is to ensure the employment of the guides also in the periods of low demand. Of course there are also other reasons for these off season tours: an improvement of Volkswagen's image in general, more press coverage in wintertime and positive effects on the work motivation of employees. Moreover, these special tours also bring along some change and challenge for the guides themselves. For example, 'Employees on Tour' is a special tour for the workers of the Volkswagen factory and their families. The tour has proven to be extremely popular among workers: 900 employees and their families had a chance to experience the VW plant in 2007 and from December 2007 to March 2008 another 900 workers with their families took advantage of this special offer. A similar tour is also available for employees from foreign branches of Volkswagen and former guest workers (from for example Italy, Tunisia, Greece and Portugal) who are offered tours in their mother tongue. Moreover, Volkswagen started to directly approach local Kindergartens and elementary schools by offering special tours for kids in the off season ('Children discover Volkswagen'). From January 2007 until April 2007, 74 factory tours with 1,523 children have been conducted. Also this special tour was very popular, which is why Volkswagen decided to extend this offer: from January 2008 to June 2008 another 3,000 kids were expected to participate in this tour.

Pricing policy Notably, Volkswagen's factory tours are free of charge (whereas the visitors of the Autostadt theme park do have to pay an entrance fee). This free of charge policy was only disrupted in the year 2006, when an entrance fee for special visitors was introduced, due to the deliberation that visitors could think: 'if it is free, it is not worth anything'. Although calculations had shown that the cost per visitor amounted to €8.70, Volkswagen decided to charge only €5 per person. However, this fee was abolished again after six months, because the time effort to handle the payments was considered too big. To give an example: VW often had troubles with foreign credit cards that were not working. As one can imagine, this is not a very good start of the tour and a rather unpleasant situation; both for the visitors themselves and for Volkswagen. If you charge a fee, this means also that you need to pay duties, which comes along with extra administrative efforts. Moreover, the people in charge realised that visitors also demand more, if they pay a fee; this puts expectations high and the firm in a weaker position. Sometimes Volkswagen has to change or even cancel tours because parts of the factory suffer a malfunction and are put out of order. In such cases, it is very difficult to explain to visitors that have already paid an entrance fee, that their tour will not take place. Such a scenario may lead to displeased or even angry visitors. All of these issues together made Volkswagen rethink its pricing policy and led to the reintroduction of free company tours.

Encountered challenges For Volkswagen Visitor Services, it was not exactly easy in the beginning to convince the other departments of Volkswagen to open up for visitors. Some departments expressed their doubts about the feasibility of visits in their production area, or simply reacted negatively. Undeniably, many objections had to be taken very serious; for example the paint shop had to be off limits for visitors, in order to ensure cleanness. However, to a certain extent the negative reactions indicate the usual initial resistance against change that occurs in almost any company. By the time some departments had opened up for visitors, Volkswagen could observe that most other departments that had reacted negatively in the beginning, had given up their objections too. Even more so, the visitor service department was actually contacted by other departments that requested to be included in the visitor tours as well. By now there is actually even a certain competition between the departments to be included in VW's factory tours. Even the people from the paint shop (off limits for cleanness reasons) have offered to admit to certain number of visitors.

Company visits challenge the safety procedures in a company, and in the case of the Volkswagen plant it was decided to prohibit visits through the plant on foot. Apart from certain areas, visitors are not allowed to step out of the visitor trains or special vehicles during the tour. Moreover, the visitor trains are only allowed to circulate on certain indicated routes. There is no age limit for kids, but children may only occupy the middle seats of the visitor trains. For the special guided tours the age limit is 10 years – simply because the special vehicles are completely open and lack side doors completely.

The maximum group size of 30 also has to do with the fact that with bigger group sizes a guide can loose track of the group more easily. Keeping control of the group is important for safety reasons, but also to ensure that people do not take photos in the factory. Admittedly, the prohibition to take photos is very difficult to control, especially since small cameras have become a standard application of mobile phones. The prohibition to take photos is not intended to circumvent espionage, but to protect the workers from 'feeling like in a zoo'. This is a very delicate issue and its importance should not be underestimated. Volkswagen had to convince the worker's union, in order to allay fears that the worker's rights could be violated. Espionage is not such an issue for factory tours, as Volkswagen simply decided to put the Research and Development department and other sensitive areas off limits for visitors.

According to the judgement of Volkswagen Visitor Services, they had admitted too many visitors to their factory in the peak year of 2001; in that year Volkswagen offered 52 tours a day. During certain periods, VW even encountered traffic jams of visitor trains in their plant, resulting in delays and problems for production. This inrush of visitors was due to the opening of the Autostadt theme park; although the Autostadt didn't even advertise the factory tours, the demand for factory tours from Autostadt visitors by far excelled the supply in that year (this was somehow a surprise for Autostadt, as they didn't expect that so many of the visitors of their theme park are actually also interested in a plant tour). After the year 2001 it was decided to reduce the maximum number of visitors and the number of tours to a somewhat lower level. Apart from this, experience had shown that there was a lower demand for tours in the wintertime, and that especially on rainy days there were a lot of tours without people. Accordingly, the number of tours in the winter months was cut back a bit.

Future opportunities Some of the encountered disturbances described above lead Volkswagen to some out of the box thinking concerning company visits: in order to be able to admit more visitors and to make the visit even more interesting and authentic, Volkswagen has plans to inaugurate an elevated walkway (the so-called production mile) that serves as a walking route through the Volkswagen plant. This walkway will run on top of all machines and internal roads and will provide great views on the production process, while not disturbing production and logistical processes at all.

Apart from this elevated walkway, Volkswagen Visitor Services is also planning to relocate its main entrance to a more attractive location within the company premises. This relocation would also increase the accessibility of the factory for visitors of the Autostadt. A task force, composed of factory management, facility management, the Autostadt and the VW Visitor Services, will work together on these plans.

Organisational aspects and budget The Department of Visitor Services is part of the Volkswagen Communications division and reports directly to Volkswagen Communications Brand and Product. This entity is responsible for leisure visitors that originate from the Autostadt, for special visitors and the Werkforum/Press Centre and has experienced a remarkable growth since 2000. Back then, only five employees were on the departments' payroll, while in 2007 the department already employed 30 people. The Department of Visitor Services is responsible for the organisation of company visits as well as press services. This is no coincidence, as Visitor Services is conceived to be pure public relations by Volkswagen. Together, these two tasks compromise Volkswagen's representative window to the public. Additionally, another 29 employees (the drivers of the panorama cars for visitors, and personnel for maintenance purposes) are related closely to the department of Visitor Services. The department of visitor services is split in three sub-entities: Coordination of special visits, Coordination of general visits and the Press centre.

The increase in employees – and in importance within the Volkswagen AG – can mainly be attributed to the increased number of requests for company tours, since Autostadt has been opened in 2000. Of course, it can also be attributed to the increasing awareness for the importance of public relations, especially connected to the former general director of Volkswagen, Ferdinand Piëch.

There are about seven to eight times more general visitors than special visitors, but this is not reflected in the organisational chart: the sub-unit that supports the general visitors only employs double the number of employees, compared to the special visitors' unit.

Apart from the functions described above, Volkswagen Visitor Services also acts as an internal and external consultant for industrial tourism. Some companies in the region have actually shown interest in Volkswagen Visitor Services: representatives from the companies that want to implement professional company visits in their firms, tend to inform themselves in detail about the management of VW's visitor tours. The Visitor Services at Volkswagen Wolfsburg also helps to implement professional company visits in the six other Volkswagen plants in Western Germany. The VW factories in Braunschweig, Salzgitter, Hannover, Emden and Kassel get substantial support by the Wolfsburg Visitor Services (both in rebuilding and redesigning facilities and even more in personal support) in order to establish a professional level for factory tours there. In general, Volkswagen prefers to open factories to the public that are characterised by state-of-the-art machines, a high level of maintenance and a high quality environment. These conditions help to give visitors a good experience. Some of the factories are currently upgrading their facilities for visitors, and in the meanwhile refer visitors to the mother plant in Wolfsburg. Also foreign Volkswagen branches, like Volkswagen Slovakia, have asked Wolfsburg Volkswagen Visitor Services for information and help. So more and more, Volkswagen Wolfsburg is recognised as forerunner in the field, and this way emerged as a partner for consulting in relation to visitor management, both in-house and externally.

The Volkswagen management grants a yearly budget to the department of visitor services, which ultimately decides how this budget is used. Volkswagen doesn't expect direct profit from the factory tours and from the department of visitor service. The conviction that the department brings profit to the company in a more indirect way (by increasing the sales) dominates and justifies the budget spent on it. Volkswagen Visitor Services admits that there is no suitable way to measure the performance of the department; that is, their true impact on the sales numbers. Nevertheless, what can be measured is the impact of their measures on the perceived image of Volkswagen, described in the paragraph below.

Reasoning behind the emphasis on company visits and evidence for success

As we have seen, Volkswagen is investing a lot of money in manpower, in its vehicle fleet and in other equipment and takes all the organisational efforts to avoid that the (up to 1,300 daily) visitors disturb the production process. On top of that, Volkswagen does not even charge the visitors for these services. The reader might ask why Volkswagen is putting so much effort in this. Well, public relations have been clearly gaining importance within the Volkswagen group; they are seen as an integral part of the pre- and post-sale services that – according to Volkswagen's philosophy – make the difference in today's competitive economy. Live and authentic factory tours are clearly seen as part of public relations and as a tool to increase the disposition of people to recommend Volkswagen, to improve the image, to increase customer retention and to increase to disposition to buy. A marketing study that was carried out by University of Mainz in 2006 to test these assumptions, confirms the existence of positive effects. In the course of that research, 244 guests were questioned before, and 242 guests after having participated in a factory tour. Almost 90 per cent would recommend the factory tour to others and the factory tour had a positive impact on every image factor that was measured (quality, dynamics, modernism, design, safety and reliability). The company tour also had a slightly positive impact on customer retention. Moreover, the data suggests that a factory tour increases the probability of another sale significantly: it increases the share of visitors that intend to buy a Volkswagen by 13 per cent – from 58.05 per cent to 65.6 per cent (Volkswagen Visitor Services 2007b). According to VW, company visits are a rather cheap and effective tool for customer retention, especially when compared to advertisements in newspapers.

Autostadt

Back in the 1994 Ferdinand Piëch was the president of Volkswagen Group. He realised the importance of public relations and envisioned the Autostadt theme park as an important tool for Volkswagens' public relations and to enhance VW's reputation. The key features of his idea were the distribution of new cars *at the source*, and the creation of a brand land: an area to display all of the sub brands of the Volkswagen Group. After two years of construction time, Autostadt opened its doors to visitors in June 2000. This project was established as a decentralised branch of the EXPO 2000 in Hannover. The EXPO can basically be seen as the spark that helped to realise the big plans of Ferdinand Piëch and as a unique possibility to launch the project within a big event. The Autostadt theme park consists of a number of modern buildings and pavilions, waterways and bridges, lakes, hills and green areas, but also of urban elements such as squares and streets. The Volkswagen Group invested approximately €430 million to set up the Autostadt on a 25 hectare area of land (Autostadt 2007c).

The reader might wonder what a theme park like a brand land has to do with the definition of industrial tourism that we gave in the introductory chapter at the beginning. Well, on the first sight this is indeed no industrial tourism attraction according to the definition. Obviously, Autostadt visitors do not learn much about the production process of the theme park itself in the course of a guided tour on the Autostadt premises. However, considering the fact that the Autostadt GmbH (limited company) is a 100 per cent subsidiary of the Volkswagen Group, and hence only represents a diminishing part of Volkswagens overall turnover, it can be considered as part of a company whose overall most important goal is the production and sale of cars. Autostadt is also the spot where customers of Volkswagen pick up their new cars and where the company tours of the neighbouring Volkswagen plant start. Hence there is a clear relation to the factory.

Attractions within Autostadt The brands of Volkswagen, Volkswagen Commercial Vehicles, Audi, Bentley, Lamborghini, Škoda and Seat operate their own pavilions, where they display their newest car models and orchestrate the image of the brand. Image improvement has been a major driver of the brand pavilion concept. Only the sub-brand of Bugatti does not have its own pavilion. Apart from the brand pavilions, there are a couple of other important buildings on the Autostadt premises. Konzernwelt makes customers familiar with the technology, design and values of the Volkswagen Group. In this building visitors have for example the opportunity to design their own car or to play a computer game that deals with the logistic process related to car production (description of the game: 'in order to assemble a vehicle, parts are needed from all over the world. By using various methods of transportation, these parts have to be dispatched around the globe in the most cost-effective way while at the same time meeting existing deadlines'). Moreover, there is a driving simulator, a cinema and a mist tunnel in which visitors can experience how limited vision affects their sense of confidence or safety. Visitors can also get information about research and development at Volkswagen, and about investor relations. This is just a selection of all the attractions in the Konzernwelt-building. Furthermore, there are also a number of attractions for kids. A neighbouring building, the Zeithaus (time house), deals with the history of the automobile in general and also shows a number of cars from competitors. For an extra fee, also the two towers that are used for the storage of cars can be visited. In Autostadt visitors can also take a ride on Volkswagens off-road circuit, they can participate in a training to learn how to drive more safely or how to drive in a more economic way. Five times per day visitors can also participate in a boat trip on the Mittellandkanal (Autostadt, 2008).

Further on, the Volkswagen power station – the main landmark of Wolfsburg – is located between the Volkswagen plant and the Autostadt. Only one half of the power plant is still in operation, the other half is operated by the Autostadt and has been completely refurbished. Today, this part of the building hosts a concert hall, where many of the events that are organised by Autostadt take place (such as Movimientos: the festival weeks of Autostadt). The power station is an example how industrial heritage can be integrated in an industrial tourism concept.

Ritz Carlton Hotel The Ritz Carlton Hotel can be considered one of the attractions and eye catchers of Autostadt. Due to the importance of hotel accommodation in relation to industrial tourism and the fact that this five star hotel is an integral feature of the Autostadt concept, we dedicate a separate section to it. This hotel is located at the edge of the Autostadt terrain, directly at a harbour basin of the Mittellandkanal, and has spectacular views on the Volkswagen power station on the other side of the harbour basin. This branch of Ritz Carlton is the first hotel of the chain which is located in the midst of an industrial area, and had also been the first Ritz Carlton Hotel in Germany (soon afterwards, Ritz Carlton opened their second German location in Berlin).

It was, however, not that easy to get this five star hotel to Wolfsburg. The Volkswagen group itself bore the investment costs for the erection of the hotel and rents the building to Ritz Carlton. Moreover, Volkswagen had to guarantee an occupancy rate of 60 per cent by the Volkswagen group itself (business guests, people that pick up cars and so on). Without this guarantee, it would not have been possible to attract the Ritz Carlton to Wolfsburg; the investment basically seemed to be too risky for them.

The hotel has a total of 174 rooms, including 21 suites and 34 club rooms. Moreover, it accommodates a restaurant that was rated with two stars by Michelin. One of the most spectacular features of the building is the outdoor swimming pool: a construction that floats in the harbour basin and which is heated constantly to 30 degrees centigrade (Ritz Carlton, 2008). A lot of the hotel's guests are actually leisure visitors. The hotel is also popular with clients that come to Wolfsburg to pick up their new cars. This last target group is actually actively encouraged to come to the Ritz Carlton Hotel – they are entitled to spend a night in the hotel for a substantially reduced price. Volkswagen tries to offer a memorable experience for the clients that pick up their new cars, and obviously they are even willing to subsidise their hotel bills.

Car pick ups Car pick ups are the most obvious and direct link of the Autostadt with industrial tourism in the way we defined it: with the opening of the Autostadt in the year 2000, Volkswagen also introduced an innovative way to attract the buyers of new VW cars to the Volkswagen homestead in Wolfsburg. Since then, people that buy a new Volkswagen car in Germany, the Netherlands or in Austria can choose whether to pick up the car at their local car dealer – as usual – or to bring the registration plates of their future car personally to Wolfsburg, participate in a factory tour, have a nice day out in the Autostadt theme park and then drive their brand new car back home. Volkswagen provides the car collectors with free entrance tickets for the Autostadt and invites them for lunch.

However, the possibility to collect one's new car exists only for buyers of a Volkswagen, not for customers of any other sub brand of the Volkswagen Group (such as for example Audi). This innovation for the delivery of new cars was preceded by tough negotiations with car dealers in the three countries, who feared to lose part of their business to a new direct selling system. Evidently, the car dealers confused Volkswagen's system of product pick-up at the source, with the more popular factory direct sale, characterised by a manufacturer selling directly to the end-user of a product. And indeed these two concepts seem to be similar on the first sight, with the notable difference that Volkswagen only distributes the car from the source, while the sale is still done by an intermediary. Hence, the fears of the distributors were for no reason, as the customers still buy their car at their local car dealer, but just pick it up somewhere else. Similar to many companies that employ a factory direct sale system, also Volkswagen gives clear incentives for picking up the product at the source: it is simply cheaper for the customer to pick up the car there, compared with the more expensive delivery to the car dealer.

The philosophy behind the car pick ups is that the customer gets a feeling 'where the baby was born', although – of course – not all Volkswagen models are produced in the Wolfsburg plant (this is by the way another crucial difference to factory direct sale, where only products that are produced in the factory itself are sold). Other VW models are basically transferred from the other plants to Wolfsburg (such as the VW Eos from Portugal, the VW Touareg from Slovakia or the New Beetle from Mexico). These cars are stored in two 48 metre-high glass towers that have a maximum capacity of 800 cars (Volkswagen Visitor Services 2007c; Architecture in Wolfsburg, 2007). The new cars arrive in the tower through a glass bridge/tunnel that connects the towers with the neighbouring Volkswagen factory. The cars are lifted up into the tower by an electronically-controlled storage system, and remain there until their new owner arrives to pick them up.

This new system of car delivery has been a success story and a way to make people more familiar with the values and ideas of the Volkswagen group: in the year 2006 roughly 170,000 customers came to Wolfsburg to pick up their new Volkswagen, by the year 2007 Volkswagen had handed over more than 1 million cars. By now around one third (!) of all people that buy a new Volkswagen in Germany, decide to pick up their new car at the Autostadt in Wolfsburg (Autostadt, 2007), so this policy can be definitely considered rather popular with the clients.

Visitor Numbers Nowadays, between 500 and 700 people come to Wolfsburg daily to pick up their new Volkswagen. This means that every 45 seconds a car handover takes place. Each visitor that picks up a car brings along 2.3 persons on average, which means that Autostadt receives between 1,150 and 1,610 visitors per day in the course of the car pick ups. If we take into account that Autostadt on average receives around 5,500 visitors per day (and even up to 20,000 at weekends), this means that 20 to 30 per cent of the Autostadt visitors come there to collect their car.

Initial studies predicted a total visitor influx of 1 million persons per year. This initial prediction was exceeded by far: In the year 2006, 2 million visitors paid a visit to the Autostadt theme park. Of these, around 400,000 were customers that collected their cars, 100,000 were pupils that participated in specific workshops and activities organised for this target group, and another 1.5 million visitors were other visitors of Autostadt or of one of the events organised there. Since the year 2000 more than 15 million visitors have paid a visit to Autostadt. The major share of these visitors actually comes from Germany: about 40 per cent of all visitors come from within a radius of 100 km. But Autostadt also draws around 9 per cent of international visitors (Autostadt 2007c). A considerable part of these 15 million visitors attended an event that took place in one of the facilities of the Autostadt (for example a concert or dance festival) and has not necessarily also visited the brand pavilions and the other main attractions that are related to cars or mobility. Hence, a large proportion of the visitors have not directly come in touch with the values and ideas of the Volkswagen group (which was one of the initial thoughts behind Autostadt). These events may increase the revenues and earnings of Autostadt, but we do not know anything about the contribution of these activities to the image of the Volkswagen group.

The Autostadt has a lot of facilities that are not fully used by the 'regular' Autostadt visitors. By organising huge events Autostadt tries to bring in additional visitors, and admittedly, could achieve this goal more than well. For the future, even more events are planned: the aim is to host around 160 events per year within the facilities of Autostadt, ranging from concerts to TV-talks, dance festivals and literature readings. Although some of these events do not have any thematic link to Volkswagen, to cars or to mobility, this can be considered a smart policy, as Autostadt reaps monetary benefits by using existing capacities to the full. The direct influence on visitors that come in for events might be smaller when compared to the 'regular' Autostadt visitors, but on the other hand these events enable Autostadt to reach people that would normally not pay a visit to the theme park; and to a certain extent, also these people are exposed to Volkswagen and might get a positive feeling about the company.

There are plans to expand and change Autostadt in the near future. The reason given for this is that a theme park needs to rejuvenate itself every now and then, in order to offer something new to visitors and to stimulate second time visits. There seems indeed to be evidence for this reasoning: Autostadt has been a beloved attraction for coach tours in its beginning; however, by now many coach tour operators have withdrawn the Autostadt tour from their offer. Autostadt realised that it had to bring in some more variation in order to be able to keep the high visitor numbers from the beginning. A number of initiatives had been set to ensure a constantly high level of visitors. Among other measures, the focus on events should ensure this. Moreover, the brand pavilions of the car brands of Volkswagen are refurbished on a regular basis, in order to ensure the attractiveness for second and third time visitors. Today, more than half of the visitors have already been to Autostadt several times.

Organisational issues and budget Autostadt employs 1,400 people and accommodates three shops and nine restaurants (Autostadt 2007b). As we have already mentioned, Autostadt GmbH is a 100 per cent subsidiary of the Volkswagen Group. Nevertheless, the management is, to a certain extent, independent from the Volkswagen Group, with the effect that the coordination between the theme park and the visitor service of the Volkswagen plant is not always satisfactory. Some of the encountered frictions also reflect personal rivalries of leading actors. However, one has to keep in mind as well that the interests of Autostadt and the sub-brand of Volkswagen sometimes differ. Volkswagen represents only one of the eight sub-brands of the Volkswagen Group, while Autostadt GmbH is intended to represent all eight sub-brands. A practical example for different interests is the solicitation of Autostadt to organise company tours in the neighbouring plant also at weekends. This request was declined by the Volkswagen board of management, because Volkswagen wants to convey an authentic image to its visitors. The only areas of the factory that are busy at weekends are the repair areas (which were not considered exciting enough for the visitors) and the press shop. It can be considered as a crucial aspect that Volkswagen's factory tours only take place during hours of operations.

Contrary to the Volkswagen factory, Autostadt is open to visitors also at weekends. The entrance ticket for one day costs €15 for adults and €6 for children. Guided tours of the park are available at extra cost. In financial terms, this also means that Autostadt yields more direct income than the Volkswagen Visitor Services. Around 70 per cent of all running costs can be covered through entrance fees, as well as through the shops and restaurants in the theme park. The remaining 30 per cent (the loss) are justified as an investment in the improvement of Volkswagen's image – which is expected to increase sales. This situation is not seen as a problem at all, as the actual yields of Autostadt are higher than initially expected.

Interestingly, also the Volkswagen car museum is managed by Autostadt GmbH. However, Autostadt does not actively promote this attraction, which had already existed long before Autostadt. This car museum is located outside the Autostadt premises, some distance away at the other side of the Mittellandkanal.

Reasoning behind the Autostadt concept and evidence for success The Volkswagen Group is not the only company that has erected a company theme park in the last couple of years: also the competitors BMW and Mercedes have recently opened similar theme parks. Thus, it seems that an entire industry is in the process of discovering the benefits of such facilities. These company theme parks are generally not intended to be pure leisure places only; the companies pursue the goal to strengthen their core businesses with the help of such a theme park. Following this way of thinking, also Autostadt is intended to reflect the key values of the Volkswagen group, which are: safety, quality, social responsibility and respect for nature. These values are, however, understood in a very broad way and do not only explicitly refer to the automobile. For example, Autostadt tries to express the key value high quality through a focus on quality and organic food in the restaurants on the Autostadt premises. The theme park offers a number of attractions that do not always explicitly refer to the Volkswagen Group – according to the press department of Autostadt, this is on purpose – the common mindset there is that most visitors come to the Autostadt for the theme park itself, and are not so much interested in Volkswagen. The neighbouring factory is only seen as an essential contribution to the Autostadt in relation to the customers that pick up their cars; a large share of them decides to combine their Autostadt visit with a factory tour. Actually the theme park stopped to explicitly communicate information about factory tours in the nearby VW plant. The reason for this decision is the limited number of places that are available at Volkswagens' factory tour, and the conviction that the factory tours are not a key theme of the Autostadt. This separation makes sense to a certain extent, as Autostadt represents the Volkswagen Group as a whole, while the neighbouring VW factory represents only one sub-brand of Volkswagen. It may be problematic for Autostadt to give a favourable position to one of the eight sub-brands it is intended to represent.

Nevertheless, it seems that visitors of the Autostadt are not fully aware of the organisational split between Autostadt and the Volkswagen brand, and that they do consider the neighbouring Volkswagen plant as a main attraction of the Autostadt. This became evident through a survey that was held shortly after the Autostadt had opened its doors. Asked for the biggest attraction of the Autostadt, many visitors named the plant visit at the neighbouring Volkswagen factory, and this although the plant tours were not among the given options of attractions on the questionnaire.

Although the Autostadt does not clearly communicate its main target groups, we learned during our visits that there is a specific focus on (1) people that come to pick up their cars, (2) women and (3) children. We have already dealt with the first group in detail earlier, so here we will focus on the reasons to put the emphasis on women and children: Autostadt tries to be appealing for female visitors simply because research has shown that many times women have considerable influence on the buying decision for cars. Also kids are expected to be decisive for this buying decision. Around 100,000 pupils per year participate in the special education programmes of Autostadt and 70 teachers are directly employed by the educational department of Autostadt. Local school classes from Lower Saxony and from neighbouring German federal states are invited to the Autostadt, where the kids can obtain for example a 'kids driving licence' or participate in workshops, for example on the theme of mobility. These educational programmes usually take one full day, but sometimes programmes are also arranged for several days. To ensure the influx of kids, Autostadt has a partnership with the Ministry of Education of Lower Saxony, and approaches schools in a pro-active way.

The theme park is well aware that this emphasis on women and kids is not a standard strategy for a car company, but there is a conviction that these target groups can be indirectly influenced to buy a Volkswagen car, or can influence other persons to buy one (or one of the other brands of the VW group). Research that was commissioned by Autostadt indicates that, per year, up to 10,000 people that had formerly driven a car from a competitor, had decided to buy a car of the Volkswagen group after their visit to Autostadt. The detailed results of this study are not available to the public. Anyhow, there are strong indications that their concept actually works.

The impact of Autostadt on the economy, spatial conditions and image of Wolfsburg Due to the opening of the Autostadt in 2000, the tourism industry entered into a steep upward trend: both the supply of tourism products and the demand from tourists have increased since then. In 2006, Wolfsburg was even among the cities in Germany that could experience the biggest increase (+6.3 per cent) in tourism arrivals. The national German average was only +3.5 per cent (Autostadt, 2007). Wolfsburg has undergone drastic change and has done nothing less than to re-invent itself: this has not gone unnoticed with the editors of travel guidebooks. In fact, Merian (a German publisher of travel magazines) has recently even issued a special guidebook about Autostadt, which of course also includes information about other sights in Wolfsburg and the region. A recent survey suggests that the extension of the touristic facilities has had an impact on Wolfsburg's image. Wolfsburg was still associated most times with Volkswagen (respectively Autostadt), but also sports (the soccer club), the art museum and the Phaeno Science Centre came in the mind of respondents.

From 1999 to 2006, hotel capacities have been expanded by around 70 per cent: from 1,630 hotel beds in the year 1999 to 2,764 eight years later (Wolfsburg Marketing, 2007). These investments were only possible due to the higher number of visitors that were attracted by Autostadt. However, in the same period we can only observe an increase of around 45 per cent in overnight stays (derived from Table 2.1). So the opening of the Autostadt in 2000 led to a marked increase in the absolute number of overnight stays, but not to the extent as was expected. An explanation for this could be the increased accessibility of Wolfsburg due to the advent of the high speed train, and a resulting diminishing need to spend the night in Wolfsburg. Another reason is that Autostadt attracts a blend of visitors that is less likely to spend the night in the city. Still today an overwhelming share of the visitors is excursionists: more than 90 per cent. However, visitors that come to Wolfsburg to pick up their new Volkswagen are more likely to stay overnight and to book a package that includes other sights in Wolfsburg.

Since the opening of the Autostadt theme park, a new footbridge connects both sides of the channel – and hereby reduces the historic separation of the city and the Volkswagen premises. The bridge was built right next to the train station (*Bahnhof*), and nowadays the Autostadt theme park is only a short walk away from the station and the city centre. Nevertheless, most Autostadt visitors never get a glimpse on the centre of Wolfsburg and hence also do not spend any money there. The visitor profile of the typical Autostadt visitor is therefore in financial terms not so beneficial for other actors in the city of Wolfsburg. Other attractions that were built after the Autostadt are expected to attract visitors with a more beneficial spending profile. For example the Phaeno Science Centre is located right next to the city centre and hence more spill-over effects are expected; also the newly established Designer Outlet Centre is, naturally, expected to draw a crowd that spends their money more eagerly.

5 Analysis

So far, this case study has given a descriptive overview of Volkswagen's industrial tourism attractions. In this section we want to relate the Volkswagen case to the frame of analysis we developed in Chapter 1, and analyse the case with the help of the framework.

Target group selection

Concerning the scope of target groups, both Volkswagen Visitor Services and the Autostadt theme park emphasise that they welcome all different kinds of visitors (no visitors are excluded). However, if we look on their actual policies we can clearly see that certain groups receive more attention than others.

The Volkswagen Visitor Services (factory tours) distinguishes between leisure visitors, who receive a rather standardised tour, and special visitors – the latter receive a tailor-made tour. Apart from journalists, researchers, politicians, importers and car dealers, also student groups from universities receive a special tour and special attention. Interestingly, also customers that complain about a malfunctioning car are invited to participate in a special tour and additionally for lunch. This special treatment aims to avoid displeased customers that communicate their anger to other people. Other special target groups are the citizens of Wolfsburg and delegations from other companies in the region.

As far as the Autostadt theme park is concerned, three special target groups can be identified: buyers of cars, women and children. Customers that come to Wolfsburg to collect their new car receive a special treatment, in order to increase customer retention and other factors. Women and children are considered to have important influence on the decision to buy a car; they are influenced in a rather indirect way, for instance through events. The focus on events, children and women indicates that the Autostadt theme park is aiming to address quite a different set of wants and needs than the company tours in the neighbouring plant. This suggests that this facility draws a different blend of people, to a certain extent at least. Autostadt officials are convinced that their visitors are different from the visitors of the neighbouring Volkswagen factory. Nevertheless, experience has shown that Autostadt visitors do show a lot of interest in participating in a factory tour. Thus, although this has not been fully realised by the Autostadt, the plant tour can be definitely seen as one of the highlights for Autostadt visitors.

Concerning the scale of target groups for both Autostadt and Volkswagen, we can clearly state that most visitors come from within a radius of 100 km. Around 9 per cent of all visitors come from outside Germany. Volkswagen and Autostadt have policies to attract local and regional visitors (for example agreements with the Ministry of Education of Lower Saxony and with the city of Wolfsburg). Apart from agreements with car dealers in the Netherlands and in Austria (also customers from these two countries can collect their new car at the Autostadt), and a special factory tour for foreign VW employees, there are no specific policies geared on foreigners.

Attractiveness of the industrial tourism product

A second important issue is the attractiveness of the visit. The product of the company (cars) is a highly attractive consumer product. Also the highly innovative and largely automated production process is very attractive for visitors. But Volkswagen does not rely on these very conducive conditions for industrial tourism: both Volkswagen Visitor Services and Autostadt are continually trying to implement innovations in order to stay attractive respectively to increase the attractiveness even further. In the case of Volkswagen Visitor Services this manifests itself in a constant effort to create new kinds of factory tours; and for Autostadt the regular refurbishment of the brand pavilions is a good indicator. While Autostadt is constantly innovating in order to stay attractive for second time visitors, Volkswagen is mainly doing this to increase its attractiveness for specific target groups, and also to avoid that the guide's work becomes monotonous. Only highly motivated guides ensure that the message you want to convey is transmitted in the intended way. This is also why Volkswagen only relies on professional guides that are employed on a full-time basis with VW.

Also the combination of a factory tour and a company theme park seems to increase the attractiveness for the visitors. We assume that a two-way-mechanism is at work here: firstly, Autostadt has a positive impact on the Volkswagen plant and secondly also the plant is an important addition to the attractions of Autostadt. The first proposition is supported by the visitor statistics of the Volkswagen plant: it is obvious that from the year 2000 onwards the Autostadt has been fuelling the visitor numbers of the neighbouring VW factory. While there were only 40,945 visitors in 1999, the plant received 262,119 visitors in 2001, only one year after the opening of Autostadt. Moreover, the positive impact is also underlined by the fact that factory tours evidently face stricter limits in terms of visitors, when compared to a theme park. The reason for this is simply that within a certain timeframe, only a certain amount of visitors can visit a company, without disturbing the production process. In this sense a theme park poses a way to reach even more people (and expose them to the values and ideas of the company), than this is possible with factory tours. Volkswagen Wolfsburg is an excellent example for this: Autostadt actually managed to draw more visitors than the neighbouring plant in the year 2006 – while 2 million people had paid a visit to the Autostadt, 'only' around 187,000 people participated in a factory tour in the neighbouring Volkswagen plant. Additionally the theme park can be understood as an attractive supportive programme for the factory visits, with entertainment, leisure and educational elements (nevertheless, as the park managed to draw about 11 times as many visitors as the car plant in 2006, it is much more than a supportive programme; it can be considered an important industrial tourism attraction in its own right). Our second proposition (the plant is an important addition to the attractions of

Autostadt) is evidenced by the fact that many Autostadt visitors are interested in visiting the neighbouring factory, and consider it as one of the main attractions of the theme park. So we assume that the possibility to participate in a factory visit simply increases the attractiveness of the Autostadt as a whole.

Accessibility is an important determinant of the attractiveness of a visit. In this sense, Volkswagen Wolfsburg seems to be in a favourable position on the first glance, as the Volkswagen factory is located close to the city centre of Wolfsburg. Nevertheless, an artificial waterway (the Mittellandkanal) cuts off the factory from the city centre. Recent investment in infrastructure (a new bridge for pedestrians) has improved the accessibility, notably of the Autostadt theme park. Notwithstanding this improvement, the number of industrial tourists (visitors of Volkswagen factory tours and/or of Autostadt) that pay a visit to the city centre is rather low. In terms of external accessibility Wolfsburg is in a good position, being a stop on the high speed trajectory to Berlin. From the train station it only takes a couple of minutes to walk to Autostadt. Recent tourism developments (the Phaeno Science Centre and the Designer Outlet Centre) are concentrated around the train station and thus, are also very close to Autostadt. This means that there are some other primary tourist attractions rather close by.

Benefits in terms of income and image for city an enterprise

As Volkswagen is the main driver of industrial tourism (and of tourism in general) in Wolfsburg, we can actually expect that industrial tourism has an impact on the income and the image of Volkswagen: income and image can be considered to be highly interrelated. Although neither Autostadt nor Volkswagen's department of Visitor Services deliver any direct profits, both organisations are convinced that the image improvement that is related to their actions in the end pays off in the form of thousands of extra cars that are sold. This conviction is also underpinned by a study, which confirms that the factory visits increase the disposition to recommend Volkswagen, improve the image, increase customer retention and the disposition to buy. Volkswagen considers company visits a rather cheap and effective tool for customer retention, especially when compared to advertisements in newspapers. Also for Autostadt there are strong indications that their concept actually works. Around 70 per cent of their expenses are covered by entrance fees. In our opinion it is necessary that Autostadt also delivers a certain amount of direct income, considering that a big number of visitors come for events (which are organised within the facilities of the Autostadt) and do not pay a visit to the Konzernwelt or the brand pavilions – hence, also the exposure to Volkswagen's key values and ideas is less strong among this group. Thus, we can expect that this group does not contribute as much to the improvement of VW's image as other groups do. On the other hand, the major part of Volkswagen's factory tours is free of charge, so this department does not deliver any considerable amount of direct income. We can state that there is a certain immediate direct income from industrial tourism that does not cover the costs, plus an additional income that can be only reaped in the medium term, and that roots in the improvement of the company's image. Apparently, Volkswagen considers the sum of monetary income and the impact on the image of the company to be positive on overall.

Concerning the impact of industrial tourism on the image and income of Wolfsburg we can state that the issue becomes a bit more delicate, but all in all the outcome can be judged to be a positive one as well. The details speak for themselves: the aims of Volkswagen Visitor Services and the city of Wolfsburg seem to overlap partly, even in the short term. Volkswagen's plant tours pay specific attention towards special visitors; this includes businessmen, journalists, researchers, and so on. Those kinds of travellers are mostly travelling on business, and are generally considered to be most important for cities, since they spend most money (these people are on expense accounts). Moreover, the Ritz Carlton Hotel, initiated by Volkswagen, clearly attracts a blend of people that spend more money than average. Also international visitors generally bring in more money and tend to spend more nights in one place. We do not have data about the share of foreigners participating in VW's plant tours, but we do know that 9 per cent of the Autostadt visitors are international. Moreover, we know that around 20 per cent of the hotel guests in Wolfsburg are foreigners. Without VW and the Autostadt, the city of Wolfsburg could possibly never attract as many foreigners, so also in this sense the outcome for Wolfsburg can be considered positive. Nevertheless, the major share of visitors of the Autostadt and the Volkswagen factory are clearly not the ones that spend a lot of money in Wolfsburg. Both sites mostly attract excursionists who do not spend the night in the city and hence spend less money than the average tourist. Thus, from a short-term income point of view only some of the visitor groups are very beneficial for the city of Wolfsburg. In our opinion, the reason for this overwhelming share of excursionists lies partly in the fact that the city of Wolfsburg and Volkswagen do not have any specific agreements to focus on common target groups. To make things worse (at least from the viewpoint of short-term income), the Wolfsburg city marketing office even directs its promotion efforts on bus tourists, which – ironically – is one of the tourist groups that tends to spend very little money. In spite of that, we acknowledge that a big amount of visitors is needed if a city wants to improve its image; if you have a lot of 'witnesses' that a place is actually not as dull as everybody thought, this message will spread more quickly. Thus, also these visitors are needed, if a city wants to improve its image.

However, Wolfsburg is well aware that other types of visitors may have a more favourable spending profile. The city itself has launched policies to attract these kinds of visitors, and one has to be aware that these policies are in a way clearly based on the success of Volkswagen's industrial tourism attractions. Namely, Wolfsburg had used the advent of Autostadt to broaden the range of available tourist attractions. Industrial tourism was thus the spark which triggered new developments in the local tourism industry. While Wolfsburg was formerly only known as the Volkswagen city, today a broad range of tourism attractions gives the city of Wolfsburg a somewhat more diversified image and more appeal for visitors.

The consequence for Wolfsburg is that the city is likely to attract more visitors with higher spending profiles in the future. Some of the new tourism developments, like the designer outlet centre, are located at the edge of the city centre – near Autostadt – and are intended to attract a more thriftless crowd. Moreover, the realisation of Autostadt helped the city to bring the tourists closer to the inner city. Before the advent of Autostadt, the more peripheral VW plant was the only crowd puller in Wolfsburg and thus tourists were mainly spending time in an area where they couldn't spend any money. Volkswagen and Wolfsburg have always been linked to one another in economic terms, so there has been – and will be – a certain overlap of economic interests between these two actors. Moreover, the connection of the city centre and Autostadt through the City Bridge has led to closer links in spatial terms, which in the end enforces also the overlap of economic interests.

Organisational set-up

As already described, Volkswagen's industrial tourism activities in Wolfsburg are characterised by an organisational split between the Visitor Services (which deals with factory tours for the Volkswagen sub-brand) and the Autostadt theme park (which is intended to convey the key values and ideas of the Volkswagen group to visitors). This internal organisational structure leads to a number of frictions and coordination problems. Pooling these two units together in one department would be very unrealistic, but it could be a good idea to establish some more (personal) links between Volkswagen Visitor Services and Autostadt; this would facilitate a common focus and direction of impact. However, as these two organisational units do not serve the same aim, a certain amount of frictions will inevitably remain.

An interesting feature of Volkswagen's Visitor Services is that it forms part of the company's Communications division. So factory visits are not seen in an isolated way, but integrated with other functions like for example press contact. Another distinctive feature of the department is its subdivision: one sub-unit is responsible for the coordination of special visits, another one for the coordination of general visits (leisure visitors). Different amounts of attention for different target groups are thus also reflected in the organisational chart of Volkswagen. While there are about seven to eight times more leisure visitors than special visitors, the sub-unit that supports the leisure visitors only employs double the number of employees. The department of visitor services experienced a remarkable growth over the past ten years, and could clearly increase its elbowroom and power within the Volkswagen AG: this department has a yearly budget and full decisive power upon it, moreover, the board of Volkswagen does not require this department to make 'direct' profit.

Also the organisational environment for industrial tourism that is provided by the city of Wolfsburg is relevant for our study. Actually, the city marketing agency of Wolfsburg does not have an explicit focus on industrial tourism, and hence there is no organisation that acts as a platform for company visits in the city or as an interface between companies and tourists. However, Wolfsburg Marketing also performs the function of a travel agency and can thus take influence on occupancy rate of for example hotels, and alleviate and steer the effects of tourism in general, and of industrial tourism in particular (by alleviating problems associated with the seasonal character of tourism, promoting visits on weekends and so on). The experience of the Wolfsburg Marketing organisation has shown that the demand for visitor packages that combine various factory visits is rather low. In Wolfsburg, such a combination is most likely requested by professionals, but not by leisure visitors. This also explains the emphasis on combinations with other, rather conventional, tourist attractions like museums. The city marketing agency of Wolfsburg plays a leading role in the promotion of these visitor packages. Interestingly, a conducive organisational framework for regional industrial tourism does not come from the city, but from Volkswagen itself. Volkswagen Visitor Services acts as a consultant in industrial tourism issues, not only internally, but also for local and regional companies.

References

Architecture in Wolfsburg 2007. *Architecture Guide.* Wolfsburg: Stadtwandel Verlag.

Autostadt 2007. Press information: *Volkswagen Unternehmensplattform erreicht neue Bestmarke.*

Autostadt 2007b. *Press Kit Autostadt.*

Autostadt 2007c. *Autostadt Press CD.* 09/2007.

Autostadt 2008. [Online]. Available at: http://www.autostadt.de [accessed: 2008].

Autouni 2008. [Online]. Available at: http://www.autouni.de [accessed: 2008].

City of Wolfsburg 2001. *Wolfsburg: A city changes its appearance: From industrial colony to service and leisure centre.* [Online]. Available at: http://www.wolfsburg-staedtebau.de/stadtwob_2.pdf [accessed: 2008].

City of Wolfsburg 2006a. *Wolfsburg Portrait.* [Online]. Available at: http://www.wolfsburg.de/stadtportrait/ [accessed: 2008].

City of Wolfsburg 2006b. *Statistisches Jahrbuch 2005–2006.* [Online]. Available at: http://www.wolfsburg.de/verwaltung/strategischeplanung/statistik/info/ [accessed: 2008].

Deutsche Bahn 2008. [Online]. Available at: http://www.bahn.de/ [accessed: 2008].

FH Braunschweig Wolfenbuettel 2008. University of Applied Sciences Braunschweig/Wolfenbuettel. [Online]. Available at: http://www.fh-wolfenbuettel.de/ [accessed: 2008].

Mader, T. (2003) *Produzierende Betriebe als touristische Attraktionen im Ruhrgebiet. Grundlagen, Erscheinungsformen, Probleme. Magisterarbeit, Heinrich-Heine-Universitaet Duesseldorf.* Hamburg: Diplomica Gmbh.

Merian 2007. *The Autostadt in Wolfsburg.*

Niedersächisches Landesamt für Statistik 2006. *Beherbergung im Reiseverkehr Tabel K 7350125.* [Online]. Available at: http://www1.nls.niedersachsen.de/Statistik/ [accessed: 2008].

Ritz Carlton 2008. [Online]. Available at: http://www.ritzcarlton.com/en/Properties/Wolfsburg/Default.htm [accessed: 2008].

Volkswagen 2007. *Facts and Figures, Navigator 2007.* [Online]. Available at: http://www.volkswagenag.com/vwag/vwcorp/info_center/en/publications/2008/03/navigator_2008.-bin.acq/qual-BinaryStorageItem.Single.File/Navigator_WEB%20engl.pdf [accessed: 2008].

Volkswagen AG 2008. *Volkswagen Company Report 2007.*

Volkswagen Visitor Services 2007. *Organisational Charts Besucherdienste Volkswagen.*

Volkswagen Visitor Services 2007b. *Visitor Services Plant Wolfsburg*, Volkswagen Communications Brand and Product.

Volkswagen Visitor Services 2007c. *Guestbook.*

Volkswagen Visitor Services 2008. [Online]. Available at: http://www.volkswagen.de/ [accessed: 2008].

Wasserstrassenkreuz 2008. [Online]. Available at: http://www.wasserstrassenkreuz.de/mittellandkanal.html [accessed: 2008].

Wolfsburg AG 2008. [Online]. Available at: http://www.wolfsburg-ag.com [accessed: 2008].

Wolfsburg Marketing 2007. *Gegenueberstellung Statistische Daten Tourismus Wolfsburg 1999–2007*, status: 15.08.2007.

Wolsburg Städtebau 2008. [Online]. Available at: http://www.wolfsburg-staedtebau.de/stadtwob_2.pdf [accessed: 2008].

Discussion partners

Mr. Nicholas A. Batten, Press department, Autostadt Wolfsburg.

Mr. Walter Fink, Head of Visitor Services, Volkswagen AG.

Ms. Karin Koenig, Special Visitor's Services Co-ordinator, Volkswagen AG.

Mr. Michael Reink, Department director city marketing, Wolfsburg Marketing.

Mr. Joachim Schingale, Department director tourism, events, congresses and fairs, Wolfsburg Marketing.

Mr. Andre Schirmer, Press Spokesman, Autostadt Wolfsburg.

Chapter 3
Industrial Tourism in Cologne

1 Introduction

In this chapter we provide a description and analysis of industrial tourism in the German city of Cologne – or Köln, as the city is called in German. We selected the case of Cologne, because of the wide offer of industry tours in different sectors. Moreover the wider Cologne region still possesses a substantial base of manufacturing companies. Many of these companies do not manufacture consumer products – with a natural appeal to visitors – but semi-final products for other firms in the industrial value chain. Other reasons to include Cologne are the city's approach to include surrounding municipalities (most notably the cities of Leverkusen and Hürth) in its efforts to promote industrial tourism, and the fact that the city has made efforts to open up public companies and organisations.

The structure of this case study is as follows. After this introduction, section 2 deals with the characteristics of Cologne, including a description of the city's tourism profile. Section 3 deals with the organisational framework for industrial tourism in Cologne and section 4 comprehends an overview of a number of industrial tourism attractions in Cologne and surroundings. Finally section 5 confronts the Cologne case with our research framework.

2 General profile of Cologne

Cologne can look back on a long history and is one of the oldest cities in Germany. In the Roman Empire Cologne – at that time 'Colonia' – was one of the most important trade and manufacturing centres north of the Alps; the city was and still is also an important religious centre. In the Middle Ages, Cologne experienced a period of prolonged wealth and later on Cologne became member of the Hanseatic League (Cologne Tourism, 2008). But the city didn't have heydays only: the centre of Cologne was destroyed severely in World War II and had to be rebuilt completely. In the post-war period Cologne established itself as the number one media capital of Germany.

In 2006 Cologne had a population of 989,766 (LDB NRW, 2007), making it the fourth biggest German city behind Berlin, Hamburg and Munich. The population of Cologne has remained relatively constant over the past 30 years, while the share of foreigners had been rising until the mid-1990s. In 2006, 31.4 per cent of Cologne's citizens had a foreign background (LDB NRW, 2007). With more than 60,000 citizens, the Turkish community is by far the single most important foreign group, followed by almost 19,000 Italians. Also the Serbian, Polish and Greek communities count more than 5,000 members.

The region of Cologne is generally considered to encompass the administrative districts of Cologne, the district of Leverkusen plus the districts of Rhein-Erft to the west of Cologne and the district Rheinisch Bergischer Kreis to the East. This 'Greater Cologne region' counts more than 2 million inhabitants. Although Cologne is the largest city of the Federal State of North Rhine Westphalia, the State Capital is neighbouring Düsseldorf (around 40 km to the north). The city of Bonn – around 40 km to the south – served as the capital of West Germany until the German reunification in 1990. Cologne is part of the densely populated Rhine-Ruhr region, a polycentric urban area that is among the biggest metropolitan areas of Europe and boasts around 12 million inhabitants.

Cologne has an excellent external accessibility with ten national highways that radiate from the highway ring encircling the city (Cologne Tourism, 2007). The city is in the exceptional position to have three international airports (Cologne-Bonn, Düsseldorf and Frankfurt) within easy reach. Cologne-Bonn Airport, only 14 km away, is the hub of the two German low-cost airlines Germanwings and Tuifly; it is Germany's second largest airport in terms of freight volume (Cologne Tourism 2007; 2008). Cologne's main train station is the intersection point of the German ICE high-speed-rail network and the Thalys high speed lines to Amsterdam, Brussels and Paris (Cologne Tourism, 2008). The station is located directly next to the Cathedral of Cologne, thus being in the extraordinary position to be located right in the heart of the inner city. The city centre is largely pedestrianised and makes the centre of Cologne very 'walkable'. There is one important restriction: the river Rhine cuts the city in two halves. This separation is overcome by eight bridges (Cologne Statistical Yearbook, 2006); moreover, one cable car and seven of the 11 existing *Stadtbahn* lines (a combination of metro and tram) link the two sides of the river. The Rhine is not only a barrier, it is also an important cargo link to other cities along Rhine and Ruhr, and provides a direct link to the port of Rotterdam. Cologne hosts the second biggest inland port in Germany, which underlines the city's function as a traffic and logistics hub.

Cologne is one of the most important university cities in Germany: more than 70,000 students (Stadt Köln, 2008b) are enrolled at the nine universities (of which five are of applied sciences) that are located within the municipal boundaries of the city (Cologne Tourism, 2008). Also other prestigious knowledge institutes, such as the German centre of air and space flight and three branches of the Max Planck Institute are located in Cologne.

The presence of the Max Planck institutes and of six branches of the well known Fraunhofer Institute in the surroundings of Cologne (Cologne Tourism, 2008) provides a strong basis for Cologne's Life Science Cluster. Life sciences in particular and the chemical and pharmaceutical clusters in general are strongly represented in the Cologne region, the Bayer Group in the neighbouring city of Leverkusen being the lead firm in the sector. Also the automotive industry is well represented here: Ford has its European headquarters and important production facilities in the city, and Citroën, Mazda and Renault are represented in the Cologne region with their German headquarters. Moreover, the electrical and engineering industry, the insurance industry, the retail sector, the construction and real estate industry, and to a lesser extent the food industry and the IT and communication industry are very important from an employment point of view, but also in terms of turnover. While all manufacturing sectors have experienced a decrease of employees over the past years, from 65,289 in 2000 to 52,118 in 2006 (Cologne Statistical Yearbook, 2006), the media sector is one of Cologne's most dynamic clusters in terms of growth. Eleven TV stations and eleven radio stations operate from Cologne and the TV stations RTL, WDR and VOX also have their headquarters in the city (Cologne Tourism, 2007 and 2008). Around 1,200 film shootings take place per year in Cologne. Although not directly visible in the table below, also the transport and logistics sector is an important pillar of the Cologne economy. The same applies to the MICE (meetings, incentives, conferences, exhibitions) industry.

The share of services in Cologne's economy is, with 80.2 per cent, exceptionally high. The secondary sector accounts for 19.5 per cent and the primary sector for only 0.3 per cent (MeineStadt.de, 2008). If we consider the number of employed people and include also the neighbouring districts of Cologne, the picture changes somewhat. In this case the importance of the primary and secondary sector together is more than 25 per cent (Cologne Tourism, 2008). Among the 434,341 people that were employed in the city of Cologne in 2006 were 208,408 incoming commuters, underlining the city's importance as an economic centre. The unemployment rate in Cologne amounted to 14.4 per cent in 2006 (Cologne Statistical Yearbook, 2006).

Table 3.1 A selection of Cologne's most important industries, 2008

Industry	Employees	Companies	Turnover in € million
Automotive	25,597	29	12,290
Pharma, Chemistry and Energy	23,619	71	16,338
Capital Goods*	45,602	392	9,598
Consumer Goods**	12,344	130	3,739
Retail***	54,644	25,420	8,840
IT and Communication***	17,057	10,844	2,056
Media and Entertainment	22,493	4,466	6,716
Insurances	32,820	90	n.a.
Construction and Real Estate	137,366	1,012	52,182

* including electrical and engineering industry; ** including food industry, *** including transport and logistics

Source: Cologne Tourism, 2008

The tourism profile of Cologne

The city of Cologne has a very strong position on the German tourism market. More than 248 hotels and other sleeping accommodations were operating in Cologne in the year 2006. These hotels provided 25,189 beds for visitors, and achieved an occupancy rate of 45.4 per cent (Cologne Statistical Yearbook, 2006). Moreover, more than 4,000 pubs and restaurants (Cologne Tourism, 2008) add value to Cologne's overall tourism product and 608 travel agencies and tour operators are in business in the city (IHK Cologne, 2008). The number of visitors[1] that spent a night in Cologne in 2007 was 2,487,251. Of these, around 30 per cent (768,240 people) were international tourists. Further on, almost 73 per cent of these foreign tourists were Europeans (Stadt Köln, 2008c). The share of foreign visitors has been growing over the past few years (by 12 per cent from 2004 to 2006). This increase is mostly attributed to Cologne's growing low-cost airport (IHK Cologne, 2007). Compared to other German cities, Cologne occupies the fourth rank when we consider the absolute number of international tourists (after Frankfurt, Munich and Düsseldorf). The total number of bed nights in official accommodations in 2007 amounted to 4,480,211 (Stadt Köln, 2008c). According to the Cologne Tourist board, around 28 per cent of these hotel nights can be attributed to trade fair visitors, around 19 per cent to congress and convention visitors, around 30 per cent to 'conventional' business travellers, and only around 23 per cent to leisure visitors. Additionally to the tourists in official accommodations, Cologne's share of so called 'couch tourists' (visitors that spend the night at a friend's place) is double the German average: an estimated 9 million in absolute terms. Moreover, estimations indicate that excursionists spend more than 100 million days in Cologne. Summing up the overnight stayers (4.48 million days) with the estimated number of couch tourists and excursionists, all visitors together spend around 115 million days in the city per year. Thus, the overnight stayers account for less than 4 per cent of all days that are spent in Cologne by visitors. These numbers can be explained by the fact that most of Cologne's visitors are not primarily there for the city's touristic attractions, but for shopping, conferences and congresses.

Cologne is known as an important shopping city, and especially the very successful Christmas markets attract shoppers from neighbouring countries and even from further away. The pedestrian street *Hohe Strasse* is actually Germany's most frequented shopping street (Magic Cities, 2008) and Cologne's status as biggest city in the area adds to the other conditions that are conducive for attracting shopping tourists. Concerning the MICE industry, the number of trade fair visitors has increased from 1,204,785 in 2005, to 1,777,320 in the year 2006 (Cologne Statistical Yearbook, 2006). In 2006, 45.7 per cent of the trade fair visitors

1 With visitors we refer to all kind of people that pay a visit to Cologne and that come from outside a radius of 30 km around the city. Tourists (or 'overnight stayers') are visitors that spend the night in a hotel, whereas excursionists are people that visit a place, but spend the night elsewhere.

(812,429 people) were professional visitors from abroad. The trade fair visitors are only partly included in the number of tourists and the overnight stays, as the visitor statistics for trade fairs includes local visitors as well. Moreover, some fair visitors may spend the night in another city nearby and hence are not included in Cologne's overnight-stay statistics. The Cologne trade fair premises cover an exhibition area of 284,000 m^2, located opposite the city centre on the right side of the river Rhine (Cologne Tourism, 2008). Here, around 70 trade fairs take place annually, drawing approximately 2 million visitors. Apart from the Cologne trade fair organisation, there are more than 50 other conference and congress facilities scattered all over the Cologne region (Cologne Tourism, 2008). Cologne continues to attract important trade fairs, such as GAMESCom to be held in 2009: a leading trade fair in the interactive entertainment industry (GAMESCom, 2008). The ten most important trade fairs in terms of visitor numbers are listed in Table 3.2.

Cologne is also an important city for leisure visitors who come to the city to experience its cultural and other amenities. The most important attraction and icon of the city is the Cologne Cathedral. This building is a UNESCO world heritage site and is the most visited tourist attraction in Germany. The touristic tradition of Cologne is very much linked with the Cathedral: already in the Middle Ages the city had established itself as one of the most important pilgrimage destinations (Cologne Tourism, 2008).

Apart from the Cathedral there are a number of spots that are important tourist attractions as well. In terms of architecture, Cologne boasts a number of modern

Table 3.2 The ten most visited trade fairs in 2006 in Cologne

Trade fair	Description	Visitors (2006)
INTERMOT Cologne	International Motorcycle and Scooter fair	173,770
Photokina	Trade fair for the entire photographic and imaging sector	162,492
IMM Cologne	International furnishing show	115,575
International Hardware Fair	Fair about tools, security, locks and fittings	61,077
Orgatec	Trade fair for office and office facilities (Furnishing, lighting, floors and acoustics)	57,189
Modellbahn	Fair for model railways	56,129
ART Cologne	International fair for modern and contemporary art	54,318
IFMA Cologne	International bicycle trade show	54,232
Entsorga-Enteco	International trade fair for waste management and environmental technology	40,375
Anuga Food Tech	International fair for food and drink technology	38,846

Source: Cologne Statistical Yearbook, 2006

buildings from well known architects, including Norman Foster, Renzo Piano, Jean Nouvel and Rem Koolhaas. Also the river Rhine is a touristic asset and is exploited as such: several operators offer boat trips on the Rhine and the Cologne Cable Car crosses the river Rhine and links the left bank with the Rheinpark on the other side; in 2007 this park was voted 'most beautiful park in Germany'. With three thermal spas within the city's boundaries, Cologne offers a combination of city and wellness tourism to its guests. Moreover, there are 36 museums in Cologne, two of which are related to industrial tourism, as they both deal with a consumer product and provide information about the production process. The first one is the Chocolate Museum, founded in 1993 on the initiative of the chairman of Stollwerk, which is a well known German chocolate brand. Today, the museum acts rather independently from the Stollwerk brand and is a self supporting body. The museum shows the history of chocolate, and provides information about modern day chocolate production (Schokoladenmuseum, 2008). The second one is the Museum of Fragrances located in the place where the world's oldest perfume factory began the production of the famous 'Eau de Cologne' perfume.

Apart from these two museums, the media sector is the most important attraction related to industrial tourism: four TV studios in Cologne (MMC studios, nobeo studios Hürth and two locations of the WDR) offer studio tours for visitors. Additionally, visitors can also attend TV shows (in the audience); there is a wide choice of famous German TV shows among which visitors can choose (Cologne Tourism, 2007). There are many other companies that are somehow engaged in industrial tourism, but they are not (yet) considered to be among the top attractions of Cologne. Various companies have opened themselves for leisure visitors, and an even greater number of companies are willing to receive delegations of professional visitors. Some of the most active companies in terms of industrial tourism are Ford, Bayer (in Leverkusen) and UPS.

In addition to the attractions mentioned above Cologne also hosts a number of events that act as crowd-pullers. The most important event is the world-famous Cologne Carnival. The carnival is actually not one single event, but consists of around 600 sub-events in the carnival period. The biggest single sub-event is the street parade on 'Monday before Lent', which attracts 1.2 to 1.9 million visitors every year. Another big event is the 'Christopher Street Day' in July, a gay and lesbian festival that is celebrated annually by more than 750,000 people in the streets of Cologne. In the pre-Christmas period Cologne organises six Christmas markets. Furthermore the Cologne region stages several major sports events, such as the Ford Cologne Marathon and the soccer games of the two renowned regional soccer teams: FC Köln and Bayer Leverkusen. Moreover, Cologne is also a big player in music events – the *Philharmonie* is just the tip of the iceberg: this internationally well known music hall boasts more than 400 events per year. Additionally, Cologne is endowed with a lively theatre scene, supported by its more than 60 theatres (Cologne Tourism, 2008).

As we can see, the tourism industry in Cologne is very versatile and the city can be described as a successful tourist city in many respects. One example of the city's success is a good hotel occupancy rate throughout the year. Although Cologne is a business city, it is not characterised by a low hotel occupancy rate on weekends. Weekends are actually fully booked when there are trade fairs at weekends, during the pre-Christmas period and during the Carnival period. There are only two periods during which the demand for hotel beds is rather low: the period after New Year until around the 10 January, and at Easter. Another indicator for success is the size of Cologne's 'touristic catchment area', which reaches well beyond the city's boundaries. Cologne is an attraction pole for visitors whose primary purpose is a visit to another place nearby. Experience has shown that even visitors that attend for example a musical show in the city of Bochum (which is around 90 km away) sometimes choose to spend the night in Cologne. Also professional visitors that come for trade fairs that are held in nearby cities like Düsseldorf and Essen sometimes decide to spend the night in Cologne. Thus, Cologne's force of attraction is noticeable for the other cities around, as the city benefits from visitors of other places nearby. This happens also the other way round: if, for example, all hotels are booked in Cologne, nearby cities receive overnight guests whose primary purpose is a visit to Cologne. Particularly during big events the hotel capacity of one city in the region is often not sufficient, so other cities nearby also benefit.

3 The organisational framework for industrial tourism in Cologne

In the previous section we already briefly touched upon the position of industrial tourism attractions in relation to other, more conventional, tourist attractions in Cologne. In the present section we turn our attention towards the twofold organisational framework for industrial tourism visits, which prevails in Cologne. While the Cologne tourist board takes care of promoting company tours for leisure visitors and student groups, the Cologne Chamber of Commerce and Industry deals with incoming professional visitors. In the paragraphs below we describe the strategies and actions of these two most relevant public bodies in the field of industrial tourism.

The Cologne Tourist Board

The Cologne Tourist Board is a public company that is a 100 per cent subsidiary of the City of Cologne. The supervisory board of the Tourist Board is made up of aldermen from the Cologne city council. Part of its budget comes directly from the city, but the organisation looks for additional funding for its activities through commercial income. The Tourist Board's ambition is to strengthen Cologne's tourism in general and to increase the added value of tourism. More in particular it aims to increase the length of visits and the spending volume per visitor. One of its tasks is to organise guided tours for tourists: about 10,000 tours a year. Although the Tourist Board explicitly represents the tourism interests of the city of Cologne only, it also cooperates with tourism organisations of two bordering districts (the Rhein-Erft Kreis and the Bergisches Land), and with the cities of Bonn and Düsseldorf.

From the Cologne Tourist Board's point of view, the contribution of industrial tourism to the overall tourism profile of Cologne is limited. Industrial tourism is not seen as a unique selling point of the city, and receives relatively little attention in the promotion of the city. Overall, more efforts are put in boosting the city's strong tourism assets than developing new assets such as industrial tourism. This also explains why there is no organisational unit within the tourism organisation that deals exclusively with industrial tourism. Besides, the promotion of industrial tourism is seen with some scepticism, as the present slogan of the Cologne tourist board is 'Cologne is a feeling', and industrial tourism in general is considered to be too sober and factual to fit with this message. As a consequence, industrial tourism is explicitly treated as a side programme to the other tourism attractions, as a nice 'add on' rather than as a proper touristic attraction. In spite of this approach, two initiatives are obviously related to company visits: the publication of a list of firms that offer company tours and an annual event called Expedition Colonia.

The publication of a list with company tours The Cologne Tourist Board displays a list of available company tours in the Cologne region on its website and in its Sales Guide, a booklet distributed among travel agencies and tour operators. The Board, however, does not take care of the organisation of company visits: it only displays information about it and forwards requests for company tours to the firms in question, acting as an intermediary. Remarkably, the list of company tours has not been put together by the Board itself, but by the publisher who printed the Sales Guide. The company tours are divided into two categories: 'Industry and Services', with the contact details of eight firms that offer tours, and 'Media and Theatre', with six suggestions. Table 3.3 comprehends an overview of this list.

Notably, only eight of the 14 institutions in the list are located within the municipal borders of Cologne; the other organisations are located in nearby suburbs of Cologne (Hürth, Brühl and Bergisch Gladbach), or even further away in the *Rhein-Neuss* district. The Cologne Tourist Board has recognised that some of the most interesting companies (notably industrial companies) are located outside the boundaries of the city. This is not seen as a problem at all, as the Tourist Board is well aware of the 'gravitational pull' the city of Cologne exercises on visitors of nearby places: the expectation is that industrial tourists that visit attractions in the surroundings of Cologne will most likely choose to spend the night in Cologne, and thus spend money in the city. A major problem with this company list is that the provided information (like telephone numbers and contact persons) is not up to date: there is no regular communication between the Cologne Tourist Board and the companies on the list. As this company list was put together already some time ago by the editor of the Cologne promotional booklet, some of the companies in the list are actually not even aware that the tourism organisation spreads information about their industrial tourism activities. Some of the companies that are included in the list are, moreover, not very active at all in industrial tourism. For example the M-Real Zanders GmbH only offers company tours for customers (leisure visitors are not granted access), and Renault Nissan Germany only organises around five company tours per year (displaying their main logistic centre in Germany and their headquarters). Furthermore, some companies that are generally considered to be very active in industrial tourism (like UPS or Bayer) are not explicitly mentioned in this list (Table 3.3).

From the list we can conclude that 'Media and Theatre' is an obviously important sector in the industrial tourism supply of Cologne. As the media sector adds a lot of value to the regional economy, visits to media companies are expected to enforce an existing strength and to have a positive impact on the image of the city. The touristic potential of the media sector has been recognised, and industrial tourism in this sector receives more attention than visits to companies in other sectors. Many people come to Cologne to attend a TV show as a spectator and/or to

Table 3.3 Organisations that provide company tours according to the Cologne Tourist Board

Category	Company	Sector	Location
Industry/Services	Ford Werke GmbH	Automotive	Cologne
Industry/Services	Renault Nissan Deutschland GmbH	Automotive	Brühl
Industry/Services	M-real Zanders fine paper production	Chemical	Bergisch Gladbach
Industry/Services	The Knapsack Chemical Park	Chemical	Hürth
Industry/Services	Rheinenergie AG	Energy	Cologne
Industry/Services	Hafen und Gueterverkehr Koeln AG	Real Estate	Cologne
Industry/Services	RWE Power AG; open cast lignite mines	Mining	Rhein Neuss district
Industry/Services	Koeln Bonn Airport	Traffic/Logistics	Cologne
Media and Theatre	Magic Media Company GmbH	Media	Hürth
Media and Theatre	Westdeutscher Rundfunk (WDR)	Media	Cologne
Media and Theatre	Nobeo studios Hürth	Media	Hürth
Media and Theatre	Koelnarena GmbH (event hall)	Media	Cologne
Media and Theatre	Theater am Dom GmbH	Theatre	Cologne
Media and Theatre	Cologne Opera and Theatre	Theatre	Cologne

Source: http://www.willkommeninkoeln.de 2008

participate in a background studio tour. However, only the latter can be considered industrial tourism according to our definition.

Expedition Colonia Another initiative related to industrial tourism is Expedition Colonia, organised in cooperation with Köln Comedy Festival GmbH, a private organisation that is commonly owned by RTL and other media companies. In this partnership, Köln Comedy Festival is responsible for the organisation of the event, while the Tourist Board grants financial support and takes care of some promotional activities. Expedition Colonia is a good illustration of the Cologne Tourist Board's approach to industrial tourism: it is promoted, though without using the term 'industrial tourism'.

Expedition Colonia takes place during one month of the year and gives people the opportunity 'to discover Cologne', by participating in a wide array of guided tours. These tours are organised in addition to the regular guided tours that take place anyway. As Köln Comedy Festival is a profit based company, they charge for the organisation of the guided tours. This results in the situation that visitors also have to pay for tours that are normally offered for free if organised by the company itself. The tours have to be booked in advance, it is not possible just to arrive on the spot and participate. Company tours are an integral component of Expedition Colonia: these tours are integrated into a main theme and are mixed with rather conventional guided tours in other areas. Around 70 per cent of the more than 220 tours that are offered during the event in 2008 can be considered as industrial tourism according to our definition. The offered tours allow visitors to visit the following locations (just to name a few): the newly constructed metro tunnel, Cologne's only organ building company, the backstage area of the Opera and various studios of Cologne's TV stations. But also a visit to the roof of the Cologne cathedral and walking tours in the city are elements of the event (Expedition Colonia, 2008). In order to make the tours more attractive and to catch more attention, some tours are given by prominent personages, like the mayor of Cologne or well known TV stars.

The event mainly attracts visitors from the nearby region and is also geared to satisfy the interest of the citizens. The whole event is actually a marketing measure that was developed in order to spur the influx of visitors in the off season; as we have already mentioned in the tourism profile of Cologne, the city has a lack of tourists during the Easter period. Thus, the first edition of Expedition Colonia in 2005 deliberately took place at Easter, in the month of April. In the following years the original idea of tackling the off season was abandoned, and the date of the event was kept in April, although the Easter period took place somewhat earlier then. In 2008, the festival took place between the 5 and 27 April (duration of 3 weeks).

In 2008, a 'spin off' of the Expedition Colonia occupied the 'Easter season time slot': the *Expedition Colonia for Kids*. The target group for this event are children from all over Germany. For this purpose, the event engaged in partnerships with a number of children magazines that are issued all over the country. In total

70 tours have been organised which are specifically tailored for children and families (Expedition Colonia, 2008b). Of these 70 guided tours, around 30 per cent can be considered as industrial tourism according to our definition. Thus, the share of industrial tourism is a lot lower in the children's programme, when compared to the Expedition Colonia for adults. Examples for tours related to industrial tourism are: a guided tour of the real estate development project Rheinauhafen, a backstage tour of a media company producing animated cartoons, a backstage tour of Cologne's children's opera, a tour at the Ford car manufacturing plant or a guided tour of a local hospital. With the focus on family and children, Cologne tries to tackle the common belief that cities are not a proper place for families to spend their free-time.

Cologne Chamber of Commerce and Industry

The Cologne Chamber of Commerce and Industry (IHK Cologne) represents the interests of around 12,000 companies in the districts (Kreise) of Cologne, Bergisches Land, and Rhein Erft. The two major functions of this organisation are to provide support to companies (like legal advice, start up assistance and education and training), and to participate in policy debates with regional actors (for example in tourism policy). Moreover, the IHK Cologne collaborates with public actors in the city of Bonn, with whom they commonly operate a regional economic development agency, the so called CBB (Cologne Bonn Business).

As we have already mentioned earlier, there is a clear separation of tasks between the Cologne Tourism Board and the IHK Cologne in the organisation and promotion of industrial tourism visits. While the Cologne Tourism Board is mainly responsible for leisure tourists and students, the IHK Cologne primarily serves business delegations (groups and individuals). According to IHK Cologne the communication between both agencies works well, and overlaps are successfully avoided. There is not a single department in the IHK Cologne that deals exclusively with industrial tourism. In fact there are various departments which are somehow involved in industrial tourism: the department for Tourism and Hospitality, the department for Industry and the department for Foreign Trade. The department for Tourism and Hospitality represents the interests of traditional tourism businesses in general, like hotels, restaurants, camping sites, travel agencies, bus- and taxi-companies, but also congress facilities (IHK Cologne, 2006). The department for Industry has played only a minor role concerning tourism so far: only the composition of a list of companies that are willing to receive delegations has been carried out by them. The Foreign Trade department is the one that is most involved in industrial tourism: they receive regular requests from foreign delegations (companies, authorities, public decision makers, and so on) to visit the city. An estimated 20 per cent of all incoming delegations also request the arrangement of company tours as a component of their visit programme; most of the other delegations do get the possibility to meet company representatives as well, but these meetings take place in the building of IHK Cologne.

The main reasons for welcoming foreign delegations are the attraction of inward investments and network building. Requests of foreign delegations are always replied on, but the amount of attention that is paid to the delegations varies. This depends on the added value that a potential investment could bring for the Cologne economy. The IHK is acting in a rather pro-active way when investments decisions are at stake and especially cities in China and Turkey have been approached pro-actively by the IHK in the past to encourage them to send a delegation. The IHK also recognises positive side-effects of the incoming delegations: increasing hotel occupancy rates, increasing use of restaurants and strengthening Cologne as a location for fairs and congresses. Moreover, the reception of delegations gives the city a possibility to present itself – most notably – to people who visit the city more than one time.

Countries of origin of professional visitors In order to understand the inflow of professional visitors to Cologne, we must first deal with a specific agreement that has been reached between the 16 Chambers of Commerce and Industry that operate in the German Federal State of North Rhine-Westphalia. The 16 Chambers have actually agreed to split up all foreign markets among them. The IHK Cologne is responsible for dealing with the markets of China, Taiwan, Vietnam, Kazakhstan and Turkey. Thus, the IHK Cologne is the first contact point for delegations from these countries, on behalf of all chambers in North-Rhine Westphalia. This split of tasks facilitates cross border network building, as it provides a 'one-stop-shop' for foreign political and economic decision makers. The majority of the delegations (and professional visitors of companies) in the Cologne region come from China and Turkey, the other three countries are represented to a lesser degree. Sometimes the IHK Cologne also gets requests for company tours from within Germany but this concerns mainly individual businessmen, and not groups, as this is the case with foreign delegations.

List of firms that offer company tours Similarly to the Cologne Tourist Board also the IHK Cologne has put together a list of companies that are willing to open their doors for visitors. In contrast to the Tourist Board, the chamber has really put together the list by itself, and did not charge a third party with this task. There are no standard requirements to be included in this list, also not concerning visitor facilities. When the IHK Cologne put together this list, they experienced that some firms didn't want to participate, or only wanted to participate in a reduced way. The main reason that was brought forward was the fear of not being able to protect company know-how. The list includes 30 companies – more than double the number of companies that is included in the list of the Tourist Board. Of these 30 companies, seven are also part of the list of the Cologne Tourist board, namely: Ford, Rheinenergie AG, RWE, Cologne-Bonn Airport, Magic Media Company GmbH, WDR and Koelnarena. Moreover, the list of the IHK Cologne also includes the Bayer AG, which was mentioned by several of our interview partners as a 'big player' in industrial tourism. Similar to the Tourist Board's company list, also the list of the chamber includes companies from outside the Cologne district. The reason for this is that the district of Bergisches Land and Rhein-Erft are, together with Cologne, the official field of activity of the IHK Cologne. The included companies are active in the fields of waste management, automotive sub-contraction, chemical industry, service industry, energy and water resources management, automotive, glass manufacturing, wholesale trade (including furniture dealing), lignite mining, polymer processing, aeronautics and astronautics, mechanical engineering, media, metal industry, food industry, public organisations, event-management, logistics and transport, publishing and printing as well as advertising and public relations. The experience of the IHK shows that the demand for company tours is especially high in the sectors of food industry, chemical industry, metal industry, media industry and furniture dealing. According to our interview partners, the companies that decided to open up for delegations don't see this as a burden, but are proud to show their achievements. Only sometimes – if the quality of the delegation is poor – the receiving company may get less enthusiastic to show visitors around and may strive to limit the duration of the visit. It has also has happened that companies want to be deleted from the list.

Organisation of the visits The organisation of company visits is considered to be a time consuming activity that is not a core task of the IHK Cologne. This leads to a rather reactive behaviour of the IHK Cologne when it comes to company visits; in most cases they only arrange industrial tourism related activities if the incoming delegations specifically ask for it. However, the Chamber does arrange appointments for company visits on behalf of the incoming delegations, and also provides transport for them. The receiving companies just have to agree on a date for a guided tour with the Chamber, and provide a guide. The guides are always provided from the companies themselves, not by the IHK Cologne. Usually one or several members of the IHK Cologne staff also accompany the delegations on the site. Although the IHK Cologne does get a lot of requests for visiting firms during the weekend, this is usually not possible, simply because most companies refuse to open their facilities on Saturdays and Sundays. Another restriction is the long waiting time that is associated with fixing an appointment for a company tour; some big companies that receive a lot of visitors have a waiting time of up to three months. Sometimes, this time is also used by the companies to perform security checks on the potential visitors.

Cologne's industrial tourism products

In the present section we shift our attention from the organisational framework for industrial tourism, provided by public organisations, to the level of single companies. As we cannot give a complete overview of all activities undertaken by individual firms, we selected some active and interesting companies in Cologne and its surroundings: Bayer AG in Leverkusen (chemical and pharmaceutical company), Rheinenergie AG (energy), Häfen und Güterverkehr Köln AG (logistics and real estate), Industrial Park Knapsack in Hürth (chemical companies), Magic Media Company in Hürth (media) and UPS located at Cologne Bonn Airport (logistics). Four of these six companies, to be introduced hereafter, are also listed by the Cologne Tourist Board and the Chamber of Commerce and Industry as attractions for industrial tourists. Although frequently mentioned as one of the active companies, Ford will not be analysed, in order to avoid too much attention to car manufacturing companies in this study (with Volkswagen in Wolfsburg and Shanghai, and Fiat in Turin). With this selection, we tried to achieve a good balance between the different sectors that are represented in Cologne. In each sub-section we describe a company's industrial tourism products, the organisation of visits, the motivation of the company to be engaged in industrial tourism and the problems associated with the issue.

Bayer

The global headquarters of the Bayer Group are located at 12 km driving distance from Cologne's city centre, in an industrial estate along the river Rhine in Leverkusen. This estate is located next to a local train station that has frequent public transport services to Cologne's central train station, and even to Düsseldorf. Bayer AG is a transnational business group that is active in the fields of health care, nutrition and high-tech material. Accordingly, Bayer is split in three subgroups: Bayer HealthCare, Bayer CropScience and Bayer MaterialScience. The headquarters of Bayer CropScience are located in Monheim, a town just north of Leverkusen. This division also operates an important production facility in the region, located in the industrial park Knapsack (south of Cologne, in Hürth). Bayer's general administration and the management of the other two sub-groups are located in Leverkusen. The Bayer Group is represented by around 350 branch offices and production plants around the world (of which ten are in Germany) and employed 106,000 people in the year 2006. The biggest share of the workforce is still employed in Europe (57,800). Bayer produces around 5,000 different products (Bayer, 2008), the most well known of which is the medicinal drug Aspirin®.

Apart from Bayer's three main subgroups, it owns also three service companies, one of which is the Currenta GmbH & Co. OHG. This service division operates three chemical parks in Germany, among which the Chempark, an industrial area that is part of the Bayer premises in Leverkusen. It is important to note that the chemical industry is not part of Bayer's main business anymore: most chemical divisions of Bayer were split off some time ago and have become completely independent companies, like Lanxess AG. This chemical company and many other firms in the field are currently located in the Chempark. Bayer is still represented in the area with a plant that packs pharmaceutical products and a production plant of Bayer MaterialScience. However, in terms of production the importance of the Leverkusen location for Bayer has decreased considerably. Currenta is actually a joint venture of Bayer and the Lanxess AG and also takes care of the public relations of the Bayer Group. Thus, this is also the organisational unit at Bayer which is responsible for industrial tourism (Bayer, 2008). Accordingly, Currenta also operates the BayKomm Communication Centre where four full time employees work. Since 1991, this centre functions as the communication platform for all Bayer branches worldwide. It is lodged in a building that is located in the Chempark, very close to the Bayer headquarters. Bayer decided to put the focus on a multimedia communication platform rather than on company tours in order to present the company to the public.

BayKomm hosts eight thematic exhibition rooms with more than 300 interactive exhibits on an overall area of more than 1,200 m^2. Six of the exhibition rooms try to clarify the link between Bayer's products and everyday consumer products that contain an intermediate product from Bayer. The exhibition tries to put Bayer's products into a broader context and does not provide any direct advertisements for Bayer's products. Interestingly, no brand names at all are mentioned in the whole exhibition; especially for Bayer's medicinal drugs this would not be possible anyway, as it is forbidden in Germany to advertise pharmaceutical products. The two remaining exhibition rooms contain information, facts and figures about the Bayer Group and about Bayer's activities related to corporate social responsibility. The whole exhibition is intended to show the diversified activities of the Bayer group and to make clear that Bayer is 'more than Aspirin'. Moreover, the exhibition should help to correct the still widespread but antiquated image that Bayer equals chemical industry. The exhibition rooms are regularly updated, which is crucial considering that Bayer operates in a 'fast moving' industry where technology advances quickly.

A visit to BayKomm is only possible with a guide. By seeking the dialogue with interested parties, Bayer is obviously trying to avoid superficial communication and convey its message more thoroughly. The tours are available in eight languages and are only given by native speakers. This is not only important from a linguistic point of view, but also from a cultural point of view: the tour guides should be familiar with the culture of the visitors, in order to avoid miscommunication. Apart from the usual tours, Bayer also offers a round tour that is specifically tailored to kids from primary schools (BayKomm, 2008). Also groups of students receive a special visit programme which includes, apart from a visit to BayKomm, presentations by Bayer employees that are experts in the issue of interest, as well as a presentation from the human resources department which informs visitors about job opportunities at Bayer. Tours for groups can be arranged on prior notice (two to three months in advance), and take place every day between 11:00 a.m. and 4:00 p.m. BayKomm is open to all kinds of visitors and does not exclude single persons from its visits. From 4.00 p.m. to 6.00 p.m. on Mondays to Fridays, and on Sundays from 12:00 a.m. to 5:00 p.m., visitors can drop by without prior notice at BayKomm. They can participate for free in a tour of the Centre: tours start every 30 minutes. BayKomm can offer a maximum of 18 guided tours a day, corresponding to around 600 daily visitors. The maximum group size is 40 persons for leisure visitors and 30 persons for professional visitors. Since 1991,

Bayer has received more than 2 million visitors at Baykomm, of which around 10 to 15 per cent were foreigners (estimation). Visitors include practically every segment of the society, but most of them are school classes and groups of students. Bayer particularly tries to attract teachers, university lecturers, doctors, nurses and representatives from various associations, as these are considered to be key people that have an impact on the other members of their respective organisations (multiplier effect).

Close to the premises of Bayer there is one the most beautiful parks of Germany (according to a poll of a German garden magazine). The 'Japanese Garden' occupies an area of 15,000 m^2 and entrance to the park is free of cost. BayKomm is embedded in this park landscape: its roof is covered by grass and merges with this garden which is characterised by exotic plants and Japanese-style bridges and arches. Actually, this park was already created in 1913, long before the Communication Centre was built (Cologne Tourism, 2008 and Stadt Leverkusen, 2008). During the flowering period the garden attracts a lot of visitors, who sometimes also choose to visit Baykomm.

Bayer carries out company tours as well; especially the headquarters of Bayer CropScience in Monheim receive many visitors. However, Bayer does not actively advertise company visits and on the website there is no information on company visits. If Bayer receives a request for a company visit, certain target groups are admitted in most cases: corporate customers, politicians, associations and also student groups. BayKomm is not involved in this: interested groups are basically referred to the company branch in question, which decides individually upon each request. Before Baykomm was created in 1991, visitors could sometimes also visit the company premises in Leverkusen by means of a bus tour. These tours have stopped, basically because the Bayer premises in Leverkusen is nowadays for the most part occupied by other companies which operate in the chemical sector. Another reason why the focus was shifted from company tours to an exhibition centre is that most of the production processes of Bayer are relatively unspectacular for the observer: processes often take place in shut systems that don't allow any kind of observation.

Rheinenergie AG

Rheinenergie AG provides drinking water, natural gas, long-distance heating and electricity, which is mainly produced in gas-fired power plants. The company employs around 3,000 people and was founded in 2002, as a merger of the two local and 100 per cent public energy providers GEW Köln AG and RGW AG. From 2002 to 2005 the company actually retained the abbreviation GEW in its company name, and was then known as GEW Rheinenergie. Today the City of Cologne holds 80 per cent of the shares of Rheinenergie AG and the German RWE group owns the other 20 per cent (Rheinenergie AG, 2008).

Industrial tourism has been an issue for many years at Rheinenergie AG. Until 2002, however, company tours were only offered to professional visitors, as there was no need to convince or retain customers since the company held a local monopoly. With the liberalisation of the German energy market in 2003/2004 things changed: the company realised that rising energy prices made consumers more critical. This caused an increasing demand for company tours, which could not be met by the existing supply of company tours. Coupled with the belief that company visits are an interesting tool to increase customer retention and to convince people with a negative attitude (and in the end to improve the company's image), this led to the decision to put more emphasis on company visits. Company tours are used by Rheinenergie to explain critical consumers why energy prices are as high as they are, showing that the company has positive intentions. In addition, the tours also provide a direct form of customer feedback. The company considers this superior to other forms of feedback, like surveys from external agencies, mainly because the feedback is derived from an intense face-to-face contact with the customers, instead of a brief telephone contact. This communication often delivers interesting insights that are communicated to the decision makers in the communications and marketing department. For example, the firm found out through the company tours that many customers were not familiar with the new name given to the company in 2002. In response to this, Rheinenergie launched a number of promotional measures to increase the awareness of the company brand.

In organisational terms the corporate communications department is in charge of industrial tourism. Three full time employees and one employee on a part-time contract take care of company tours, besides several other tasks. Rheinenergie is mainly visited by groups, but also admits single visitors, which are normally joined with the group-visitors. The maximum number of visitors per guide is 40 persons. Visitor numbers have been clearly increasing over the past few years. In 2007 a total of 235 company tours were held, giving around 4,000 adults and 2,200 kids and teenagers the opportunity to visit the company. Around one third (estimation) of the adult visitors were professional visitors, the rest consisted mainly of interested customers of Rheinenergie. Only a minor share of visitors were foreigners, most of which came from Asian countries, Russia and African countries, whereby Asia clearly dominated. Foreign visitors need to bring their own translator, as Rheinenergie only offers tours in German. Rheinenergie cooperates with Expedition Colonia and with some local schools, adult education centres, vocational schools and universities of applied sciences, which have integrated a Rheinenergie tour into their teaching curricula. Furthermore, the firm participated in the 'Cologne Theatre Night' (2007) and put some of its corporate buildings at the disposal of this event. Although the visitors of this event didn't take part in a company tour (and hence are not included in the official visitor numbers), they did get in touch with the company. This way, it was possible to attract visitors that are not customers of the company and make them aware of the brand and the company tours.

Tours are provided free of charge for all four sites that are open for visitors: the two waterworks, the long-distance heating tunnel and a gas-fired power plant.

Except the tunnel, all of these facilities have a kind of lecturing room where presentations about the facility in question can be given. Rheinenergie AG operates 11 waterworks in the Cologne region (Rheinenergie AG, 2008b), of which the waterworks Severin and Weiler are open to public. In 2007 more than 100 guided tours took place at the Severin waterworks, whereas only about 20 groups visited the plant in Weiler.

The Severin waterworks are located just south of the city centre and are accommodated in an historic building that dates back to 1800. In the 1950s the historic machines were dismantled and from then on constantly replaced by modern equipment. Nevertheless, experience has shown that visitors are often interested in the historic evolution of production techniques, and not only in the present day production process. With hindsight, Rheinenergie recognises that it would have been better to keep and restore the old machines; this could have further increased the attractiveness of the tour. A tour at the Severin waterworks takes between 2 and 2½ hours and involves a 45-minute presentation, a visit to the plant and to the subterranean reservoir. The reservoir is a kind of columned underground hall that was erected in 1899 and is still in use today. Usually, visitors can only take a look inside the reservoir through a small window. But every ten years this underground reservoir needs to undergo complete cleaning: then the water is drained and during a period of about three months visitors can also access the historic column hall of the reservoir. The acoustics of the reservoir is said to be very good, which is why Rheinenergie organised several concerts inside this historic industrial facility. These events have attracted a lot of interest of the media.

Furthermore, Rheinenergie's gas-fired power plant in the district of Köln-Niehl has received six guided tours in 2007. This plant marks a turning point in the company's industrial tourism policy. When the plant was constructed in 2005 visitor facilities were not taken into account. After the construction was finished, the decision was made to install an additional room, dedicated to the reception of visitors. This posterior change proved to be extremely costly, which is why the decision was made that in the future, visitor facilities will be taken into account already in the original planning for every new plant.

The already mentioned 461-metre long-distance heating tunnel (Rheinenergie AG, 2008c) is another highlight in the portfolio of Rheinenergie AG. In 2007, around 100 guided tours took place here. This tunnel starts in the city centre of Cologne and runs below the river Rhine, and herewith connects the long-distance heating networks on each side of the river. Guided tours start in a small exhibition room at the beginning of the tunnel, where a ten-minute movie about the construction of the tunnel is shown. After that visitors cross the tunnel on foot and emerge again on the other side of Rhine. The whole tour takes around 1½ hours. Although the production process of 'long distance heating' is less attractive than the other processes (water and electricity production and distribution), the tunnel does attract a relatively high number of visitors. For one thing because it offers the experience of crossing the river through an underground passage, and for another because of its favourable location in the centre. In general, it seems that plants near the centre (the tunnel and the Severin waterworks) attract more visitors than peripherally located plants that are hard to reach by public transport such as the gas-fired power plant in Niehl. Probably visitors prefer attractions in the centre because they can be combined with sightseeing and dining.

As a rule, company tours to Rheinenergie are provided free of charge, also explaining why the company does not cooperate with commercial travel agencies. During Expedition Colonia visitors do have to pay, but not to the company, but to the public intermediary organisation (covering their costs). Also a local adult education centre provides tours to paying customers. Understandably Rheinenergie keeps the tours of paying and non-paying visitors separated.

HGK, Häfen und Güterverkehr Köln AG

The Häfen und Güterverkehr Köln AG (HGK) employs more than 600 people and is a 100 per cent public company: 54.5 per cent is owned by the Stadtwerke Köln (public utility company of Cologne), the municipality of Cologne owns 39.2 per cent and the Rhein-Erft district 6.3 per cent (HGK, 2008). HGK owns the second biggest inland port in Germany and at the same time is the biggest non-state owned railway company in Germany. Apart from these two core business areas, the company also has an infrastructure and real estate division, which has around 30 employees. This division takes care of the maintenance of port areas, rolling stock, 1,300 km of rail-tracks, 69 bridges, and several cargo train stations, buildings and other infrastructure owned by the company (mainly in port areas and next to rail-tracks). HGK is involved in three large real estate developments: 1) the development of a new harbour basin in the port of Godorf, 2) the development of a new terminal for the transhipment between road and rail and 3) the development of a new city area on the spot of the Rheinauhafen, the site of Cologne's former goods traffic port. The real estate division is also involved in industrial tourism: tourists can pay a visit to the port of Godorf and they can witness the redevelopment process of the Rheinauhafen.

The port of Godorf receives around 1,000 visitors a year. Enquiries for tours exclusively come from people who live in the region: groups of pupils, students or people from citizen's initiatives. Visitors are mainly interested in the way the HGK cares for the protection of nature and animals on the site, and how the plan will be put into practice. But visitors are also interested in gazing at incoming barges and at the barges on the quayside. The main reason for the HGK to show the excavation of the harbour basin to the public is to signal that they are acting in a socially and environmentally responsible way. Based on previous experience, HGK expects that visitor interest will fade away once the construction of the harbour basin is finished.

The demand for visits to the Rheinauhafen project is higher than the demand for tours to the port of Godorf. Moreover, visitors are expected to keep coming when the construction works are finished. The Rheinauhafen project is of a far bigger scale than the redevelopment at Godorf. The centrally located Rheinauhafen has become one of Cologne's (and Germany's) top property locations, evidenced by the fact that software companies Microsoft and Electronic Arts decided to locate their offices there. This former port area dates back to 1898 and was not suitable anymore for the handling of containers. This is why the decision was made to abandon this port and redevelop it into a mixed-use city quarter on the riverside, for living, businesses, offices, art, gastronomy, hotels, and so on. The new quarter is planned to be ready by 2009 (Cologne Tourism and HGK, 2008).

The decision to open the redevelopment project for visitors was, however, not taken by HGK itself, but rather by one of the company's owners: the City of Cologne. The municipality wanted to use Rheinauhafen as a showcase project, with guided tours as instruments to highlight the advantages of Cologne as a (business) location in a better way. Expectations are that the project can support the ongoing structural change, by creating about 20,000 jobs, notably in the service sector. For HGK the redevelopment of the Rheinauhafen is not considered part of their core business (logistics) and this even more applies to the provision of guided tours.

Construction works began in June 2002. In September 2003 an information pavilion about the planning of the project opened its doors. Until now, 45,000 people – around 10,000 people per year – have visited this pavilion. The guided tours of the area (free of charge) take around 2½ hours and are held by the head of the project development department himself. Such a tour includes an explanation of the Rheinauhafen concept and of every existing building (including an elevator ride in one of the crane houses), as well as a visit to the information pavilion and to the building site itself. Usually, a coffee break is held at a café next to the building site. At the end of the tour, the groups pay a visit to the German Sports and Olympics Museum, which is also located on the site. Due to the high interest in this tour, the decision was taken to cooperate with an external guide from October 2007 onwards. Since then, HGK takes care only of special interest visitors groups and of professional delegations (still free of charge), while the external tour guide, who is an architect in her normal profession, carries out the tours for leisure visitors. Leisure visitors have to pay a fee of €10 per person, of which €9 is retained by the guide, and only €1 flows back to HGK. It is planned to continue the guided tours of the area after the completion of this redevelopment project.

The Rheinauhafen draws visitors from all over the world: around one quarter of the visitors come from abroad. Approximately 15 per cent come from other European countries, whereby the biggest share of visitors comes from the Netherlands. Around 10 per cent of all visitors come from Asia: China, Vietnam, Korea, the Caucasus, India and also Uzbekistan. There haven't been any visitors from America or Australia, and only very few groups from African countries. The visitors from European countries differ considerably from Asian visitors. While most European visitors are architects, investors, project developers or civil servants, the main part of the Asian visitors are architects or leisure tourists. The composition of German visitors is very diverse and ranges from leisure visitors, to representatives of newspapers, to groups from universities and other organisations. The interest from German universities in the field of architecture has been extremely high. Remarkably, only very few leisure tourists from other European countries have been registered as visitors.

The attractiveness of Rheinauhafen for visitors is based on three features: the organisational concept of the project (with one land owner who takes care of public space and the image of the area), architecture (combining historic buildings with modern ones), and the location (an attractive waterside location at ten minutes walking from the Cologne Cathedral). The accessibility of the site has been further improved by a newly built embarkation point for passenger boats.

The Chemical Industrial Park Knapsack

The Chemical Industrial Park Knapsack is located 12 km southwest of Cologne's city centre, in the municipality of Hürth. The Park has a surface of 160 hectares and consists of two, spatially detached, areas (InfraServ Knapsack, 2008). The highways A1 and A4 are only 4 km away; but reaching the Park by public transport from Cologne is rather difficult. This Park had formerly been occupied by the Hoechst Group. After the industrial disruption of this chemical company, the park became an independent location (in 1997), which opened up for various companies from the chemical sector. Today twelve companies are located on the premises, making Chemical Industrial Park Knapsack the biggest employer in the town of Hürth. Eight of these twelve companies (among which Bayer CropScience AG) are companies that manufacture intermediary chemical products. The other companies include an energy firm that operates a gas and steam driven turbine power station in Knapsack (Statkraft), a specialist provider for bulk goods logistics and a bank. Together, these companies generated a turnover of €920 million in 2007 and employed around 2,200 employees, of which 750 were employed by InfraServ Knapsack GmbH. This company is the successor of the Hoechst group, and owns and operates the Park as well as most infrastructural facilities on the premises. If a company decides to locate on the Park, it is obliged to accept InfraServ Knapsack as a service provider in safety and communication issues. A number of other services are also available on an optional basis, such as the provision of planning, construction, engineering and maintenance services. Moreover, InfraServ Knapsack owns the Rhein-Erft Akademie (also located in the industrial park), a vocational school that provides applied education in various fields: the chemical industry, IT, and commercial and technical professions (InfraServ Knapsack, 2007; 2008).

InfraServ Knapsack has a communication department, which consists of two employees. This department deals for example with public relations in the case of an emergency, but it has recently also taken initiatives to expand the activities in industrial tourism. Until 2007 the strategy of InfraServ had been rather re-active: tours of the Chemical Industrial Park Knapsack took only place when interested groups approached InfraServ with such a request. In 2007 only five groups visited the company (spending several days in the industrial park). Recently, however, the strategy has changed: nowadays the operator of the park is working on concepts to become more pro-active, and to invite people to participate in company tours and educational programmes. This has not yet been translated into concrete actions, but as a first step to open up the park further, InfraServ has agreed to participate in the 2008 programme of Expedition Colonia.

Furthermore, InfraServ Knapsack has started an 'idea competition' among students in order to increase the attractiveness of the area and to develop a concept for industrial tourism. This competition can be considered a result of the Regional 2010 Programme, a regional development initiative of the federal state of North Rhine-Westphalia which grants subsidies for companies that decide to adopt an idea that has been developed by students. Several students from nearby universities (Cologne, Wuppertal and others) are participating in this competition.

One of the ideas that emerged is to construct an elevated walkway made of Plexiglas tubes, including several 'viewpoints' on the industrial park. These tubes should literally connect the past with the present. This thematic walk about '100 years of industrial production in Knapsack' could start at the remnants of a coal mine that was abandoned 100 years, located within the park. The thematic walk could end at Knapsack's modern chemical factories, and could include the two waste treatment plants on the premises. Particularly interesting is the new gas and steam plant that produces energy out of waste, which has recently been erected by Statkraft. The communication department of InfraServ has recognised 'energy' as one of the themes of the future, so this power plant is expected to be one of the main industrial attractions of the chemical industrial park Knapsack.

At this time, visits of the single companies are possible occasionally, but visitors have to arrange these visits directly with the company in question. For the future, InfraServ Knapsack plans to reach an agreement with every company on the premises to allow for visitors inside the plants on a regular basis. This should not only include the permission for visits, but also the provision of guest speakers from the companies' headquarters, for visiting groups with interest in a certain topic. InfraServ also plans to provide bus transport for the visitors, and to invite visitors for lunch.

The main reason for opening up is the conviction that industrial tourism can make the recruitment of skilled employees easier. Actually, most chemical companies in the region experience a lack of skilled labour. By providing company tours, the Chemical Industrial Park Knapsack wants to show its attractiveness as an employer and tries to distinguish itself from other parks nearby. Therefore, universities and schools have been identified as main target groups for company visits. InfraServ Knapsack not only wants to attract the very scarce engineers, they also expect positive effects on the recruitment of other kind of staff. Also in areas where there is no personnel shortage, industrial tourism is expected to create conditions that are conducive for the recruitment of high quality staff. Other kinds of visitors (like leisure tourists) are not an explicit target group of the Park. This is mainly because InfraServ Knapsack doesn't want to organise tours for the sake of public relations only. Of course, they also want to have some kind of 'output', and leisure visitors as such are not expected to bring major benefits to them. However, the participation in Expedition Colonia shows that InfraServ Knapsack is disposed to receive leisure visitors in the course of an event. Moreover, the company also welcomes residents and leisure visitors during the 'open days' event of the municipality of Hürth, which takes place every three years. During this event the company tours are supplemented by an exhibition that shows the links of Knapsack's products with consumer products. Experience has shown that visitors meet this exhibition with great interest.

Magic Media Company GmbH (MMC)

The Magic Media Company GmbH (MMC) was founded in 1991 and has become Europe's largest film and TV studio operator. MMC is basically an independent media centre which rents studios, storage and office space and provides studio equipment, technical services and consulting services for the film and TV industry. Through a sub-company (MMC Independent GmbH) MMC also co-produces films and TV productions and provides financial assistance for TV projects (MMC, 2008). The company operates 35 TV studios on its two sites in the Cologne region: the 'Coloneum' in Cologne-Ossendorf and the 'Campus' in the nearby town of Hürth (MMC, 2008). MMC employs 150 full-time employees and co-operates with around 500 freelancers.

In organisational terms, industrial tourism is not very visible within MMC: there is no department that deals exclusively with industrial tourism, and also the (only) person that is in charge of organising the visits (making appointments and planning) has other, more important responsibilities like the management of TV shows. At the moment MMC employs around 15 guides, most of them being students from nearby universities who combine the tasks of a tour guide with the task of a hostess (giving company to film and TV stars). To become a guide, students must first have working experience as hostess thus knowing the company premises very well. External guides are not allowed on the premises. The selection criteria for hostesses are very much based on the average wishes of TV stars: only good looking persons with good communication skills are recruited for the job. Moreover, TV stars seem to show a clear preference for females: there is only one man among the 20 hostesses that are employed at MMC. There are also procedures for the education of the tour guides: a handout provides them with the main information needed and, moreover, every prospective tour guide gets help from two established guides in order to 'learn the trade' properly. However, every tour guide has a certain amount of freedom concerning the content of the company tour, ensuring hereby a certain 'personal touch' which is considered to be very important at MMC.

MMC has a defensive approach to company visits. They started to offer studio tours in the mid-1990s, four years after the establishment of the company. The reason for this was an increasing number of requests from school classes. However, MMC does not advertise its tours, apart from the information provided on their own website. But in fact studio tours and 'television' in general seem to be such attractive products that others advertise their product for free. The Cologne Tourism Board and the Chamber of Commerce and Industry display information about the tours on their websites. Moreover, a travel agency in Cologne and another agency in neighbouring Belgium send visitors on a regular basis. Around 80 per cent of all visitors are school classes; the other 20 per cent are composed by groups from companies that are on a team building trip and by visiting groups from all different kinds of organisations and associations. Visitors come from all over Germany but also from Belgium and the Netherlands. Leisure is clearly the main motive of most visitors. MMC also receives professional visitors, whereby most of these visitors are clients of MMC. However, the number of professional visitors is rather insignificant in comparison with the amount of leisure visitors. Another department takes care of the professional visitors, and usually a project manager of MMC himself guides these groups around. For studio tours MMC does not cooperate with commercial ticket offices, although this could be an option in the future if it reduces costs and only guides from MMC are used.

At the moment company tours take place only at the Hürth location of the company. This has to do partly with the fact that Hürth is the main location of the company and that it is also closer to, and better accessible from the centre of Cologne, when compared to the Ossendorf location. The regional train station Hürth-Kalscheuren is located right next to the premises of MMC. However, the main reason why no tours take place in Ossendorf is the fact that it is more difficult to allow for visitors there. As it is the main location for soap opera productions and movies there is less space for visitors: the risk that people snatch items from the scenery is too high for MMC. Moreover the film-sceneries for the soap operas are usually dismantled right after the shooting and so there are not so many interesting things left for visitors outside the film shootings times. The studio space is too expensive to keep the sceneries there on a permanent basis. Understandably MMC does get a lot of requests of people who want to experience the sceneries of soap operas or movies that were made there. In contrast to Ossendorf, the Hürth location hosts mainly talk shows and regular productions with sceneries that are there on a rather permanent basis. In 2007 about 10,000 people participated in a studio tour at MMC in Hürth.

In the early days of industrial tourism at MMC, studio tours at the Campus in Hürth were offered for free. Because of the ever increasing demand from school-classes, the overall expenditure on company tours had increased – this is why MMC made the decision to charge for the visits. A normal studio tour has a duration of one hour and costs €7.40 per person, with a discount of €2 for pupils, disabled persons and groups bigger than 40 persons. MMC also offers a Studiotour Deluxe: an extended tour of 1.5 hours which also includes a visit to the control room. With a price of €17.50 this tour is considerably more expensive, mainly because a skilled engineer is needed to explain the control room, and the technician's salary is a lot higher than the one of a normal tour guide. At additional cost, visitors can also book a reception with champagne (€4.50/person), snacks (€6.80/person) or coffee and cake (€6.80/person); but these extras are only booked occasionally. The list of prices demonstrates that MMC intends to cover the costs incurred. The company actually earns money with the company tours. However, the profit is rather small, notably in comparison with the profit made with studio productions. Studio tours are seen as side-business, partly explaining why the company doesn't provide any resources to invest in visitor facilities or to professionalise the studio tours further.

Participants of a studio tour are informed about all necessary steps to run a TV production: how to build the decoration, how to start a production and how the sound and light systems work in different studios. Visitors are also given the opportunity to ask whatever question comes to their mind. One guide looks after a maximum of 40 visitors. MMC cannot show around more than three groups at the same time, resulting in a maximum capacity of 120 visitors at one time. A maximum of eight tours are carried out per day. The minimum number of visitors per tour is 15 persons, in order to make a profit. Studio tours take place seven days a week at MMC; whereby Sunday is usually the most frequented day. This is because no TV productions at all take place on Sundays, and hence MMC doesn't have to attune the schedule of the tours to the schedule of the TV productions. The demand for company tours varies a lot, depending on the time of the year. Demand is particularly high during May and September because these are the months for field trips of schools. During these months, however, the capacity for studio tours is limited because many TV shows are produced at the same time. This is not a coincidence: TV studios have adapted their production schedule to the inrush of visitors (school classes included), as it has become a general problem of the industry to attract enough people as live audience for TV shows. This leads to the situation that the demand for studio tours is highest during the period in which the studios are already extremely busy with TV productions.

A studio tour can also be combined with a visit to a TV show in one of the studios of MMC, and in this case the studio tour is held just before the recording of a TV show. In fact, such a combined ticket is the only possibility for single persons to participate in a studio tour, apart from the five studio tours per year that are offered through the Expedition Colonia. MMC owns the studios, but most of the productions are actually done by companies that rent studios from MMC. This also implies that MMC cannot decide on its own if visitors can be allowed to a TV show. For this sake, the company has made agreements with its four most important clients: Mediabolo, TV Ticket Service, Endemol and Grundy Light Entertainment. Due to the fact that these companies have problems to attract audience to their TV shows – especially for new and relatively unknown TV shows – it was relatively easy to reach these agreements. In this way the studio tours are used to boost the inflow of live audience into TV shows.

UPS/Cologne Bonn Airport

UPS is an American logistics company with headquarters in Atlanta. The company has operations worldwide and is the largest express-delivery and parcel delivery service firm of the world. Moreover, they offer specialised forwarding services, as well as logistics- capital- and e-commerce services. The European headquarters of UPS are located in Brussels, but since 1986 the Cologne Bonn Airport is the main logistic hub for the company's activities in Europe (UPS, 2008), employing around 1,700 people. With its more than 260 freight aeroplanes, UPS was actually the 9th largest airline worldwide in 2006, while in that same year the Cologne Bonn airport was the third biggest cargo hub of UPS worldwide (Airliners.de, 2006). The Cologne Bonn airfield is also the second biggest freight airport in Germany, with a volume of cargo of 698,000 tons in the year 2006 (Bonn Region, 2008).

The visit of UPS (United Parcel Service Inc.) takes place by night and is generally considered to be one of the highlights of industrial tourism in Cologne. While the airport is busy with passenger flights during the day, during the night several thousand cargo shipments are handled (Bonn Region, 2008). The main cargo hall of UPS can automatically handle 110,000 parcels and documents per hour (Airliners.de, 2006). In the cargo area of UPS, visitors are witnesses to the speedy handling of parcels (Cologne Tourism, 2008b) and can experience the logistic chain that is operated by UPS – unloading the aeroplane, the automatic scanning and sorting of the parcels, and the transport of the parcels on the conveyer belts (Airliners.de, 2006). Visitors need to register at least 48 hours ahead, and need carry along their passports during the visit. The minimum age for a visit is 16 years. These visits can be arranged directly through UPS, but sometimes the Cologne Bonn Airport combines a night visit to its own facilities (the two terminals, the airfield and sometimes also the maintenance area for aeroplanes) with a visit to the freight halls of UPS. The combined tour with the airport takes up to four hours. The Cologne Bonn Airport itself also offers 'normal' guided tours four times a day from Monday to Friday, excluding a visit to UPS. Because of security reasons it is not possible to visit the Tower, neither to see an aeroplane from the inside (Cologne Bonn Airport, 2008). The airport of course has excellent transport connections, including an ICE connection from Cologne Central station: also individual visitors can reach this hotspot of industrial tourism with ease.

4 Applying our frame of analysis

So far, we have dealt with the organisational framework of industrial tourism in Cologne, and with a couple of active companies in the realm of industrial tourism. Hereafter, we will provide an analysis of the Cologne case, resulting in some conclusions. This analysis will be based on the research framework that we developed in Chapter 1.

Target group selection

As the city of Cologne does not have any explicit strategy for industrial tourism, we can only base our judgement on the city's *scope* of target groups on some concrete actions that have been set in the past. Based on the public policies on industrial tourism in Cologne, we can state that the main target groups for industrial tourism are professional visitors and investors (main target group of the IHK) as well as families and children (main target group of Expedition Colonia). In our opinion trade fair visitors are the target group with most potential in Cologne, but so far a rather reactive behaviour towards this group prevails. For the single companies, the picture gets somewhat more diverse. Almost all companies have a special focus on professional visitors, like corporate customers and suppliers. Additionally most companies have other specific target groups to which they pay more attention. Bayer particularly tries to attract so called 'multipliers' such as teachers, university lecturers, doctors, nurses and representatives from various associations. These are considered to be key people that have an impact on the other members of their respective organisations (multiplier effect). The industrial park Knapsack has identified groups from universities and schools as main target groups for company visits, mainly because of the lack of skilled labour in chemical industry. For the Magic Media Company, it is not so relevant who comes to visit them (as they are in the TV business, practically everybody is their customer). What is more relevant, is that the visitors come in groups of people; groups of people allow them to 'fill up' the stands for TV productions. The Rheinenergie AG has a special focus on its existing customers, and in particular on customers that hold a negative opinion about the company. On the other hand the HGK does not have any particular target group, which might have to do with the fact that it was not the management's own decision to open up (the public owner of the company induced this decision). Thus, the key target groups of the companies mostly reflect the composition of the company's customers, as well as people that could be potential employees or, in the case of Bayer, key persons ('multipliers') that grant access to those key target groups from which a company expects some kind of benefit. Interestingly, the key target groups of the city and the companies that we observed do not exactly match. Especially students from universities do not receive much attention from the city.

Most of the companies' target groups are the citizens of Cologne and the Cologne region; only in the case of Bayer visitors from all over Germany are targeted. The recruitment of potential employees is one reason why companies focus on their immediate surroundings. Also smaller companies with a regional customer base will most likely focus on local and regional visitors. Bigger companies like Bayer also try to attract visitors nationwide, or even international visitors. Again, there are some differences between the city and companies concerning the *scale* of target groups. For the city of Cologne, the international dimension is much more important than for the mentioned companies – especially professional visitors from abroad (investors, foreign delegations) are an important group. The Expedition Colonia on the other hand tries to attract children and families Germany-wide, also demonstrating a clearly wider scope than the investigated companies in Cologne (apart from the Bayer Group).

Attractiveness of the industrial tourism product

Most of the observed companies in Cologne pay attention to the attractiveness of the visit – these companies introduced a number of standards and limits which are necessary to keep the quality of guided tours on a certain level. One important issue here is the maximum number of visitors per guide. Three of the surveyed companies (MMC, Bayer and Rheinenergie AG) handle a maximum group size of 40 people. At Bayer, this limit is only relevant for leisure visitors, whereas professional groups are never bigger than 30 people; this is to ensure that a certain amount of two-way communication is possible, which is necessary for an attractive tour. The Bayer Group offers some further lessons for increasing the attractiveness of the visit: the Bayer exhibition centre is located in an attractive park landscape, which in itself is a tourist attraction. The combination with other kinds of attractions is an interesting tool to make a company also appealing for visitor groups that would normally not visit an industrial attraction. Moreover, both Bayer and the Chemical Industrial Park Knapsack try to make their intermediary products more appealing for leisure visitors; namely by means of an exhibition that points out the links between their B2B products and everyday consumer products which contain this intermediate product. UPS has found another possibility to make its tours more attractive: they offer tours by night, which as such is already something special. Moreover, tours at night-time are very attractive for the working population, for whom it is more difficult to participate in guided tours during the day. However, night tours might not be suitable for every company. It should be only considered as an option, if a company runs its production process also during the night, so that the authenticity of the visiting experience is ensured. Apart from night tours, also the introduction of prominent tour guides (film stars, the mayor) on certain occasions (as done during the Expedition Colonia festival) can make company tours more attractive. The Rheinenergie AG gets many requests from company tour participants to explain also their historic production processes, and from this point of view it seems obvious that a combination of modern production processes with an exhibition of old machines could increase the attractiveness of a visit. Furthermore, Rheinenergie's initiative to open a difficult-to-access area to the public (the historic reservoir) every few years and combine this with a number of events, could also be an interesting tool for other companies to stimulate second time visits and to become more attractive for certain visitor groups. Another best practice from Rheinenergie is to equip existing buildings with visitor facilities, and to take visitor facilities already into account in the planning phase of new plants. The example of HGK shows that also an interesting organisational concept, unique architecture and the location of the site play a crucial role for the attractiveness for visitors.

Clearly, the accessibility of a location contributes to the attractiveness of an industrial tourism visit. The experience of the Rheinenergie AG underlines this observation; moreover, it shows that a central location and an attractive experience (walking in a tunnel below the river Rhine) can compensate for a rather unattractive product. Rheinenergie's product 'long-distance heating' is basically invisible and not especially attractive when compared to the other products of the company (electricity and water). Apart from this tunnel, Rheinenergie has also opened up two of its waterworks to the public. Of the two, the plant with the more central location receives about five times more tour-requests. A central location has proven to be crucial not only for an easy journey to the destination, but is also favourable considering that many visitors prefer to combine industrial tourism with sightseeing and the consumption of gastronomic delights (this is the experience of the Rheinenergie AG). In the case of Cologne, some companies that are located somewhat outside the city are characterised by an outstanding accessibility by public transport, which facilitates the access for single visitors. There are regional train stations with frequent connections next to the Bayer premises in Leverkusen and next to the premises of the Magic Media Company in Hürth. As a consequence, these companies are more attractive for visitors who do not possess their own mode of transport (like scholars and students). Moreover, the location of UPS at Cologne/Bonn Airport is extremely favourable, as airports in general are connected very well to their surroundings. Of all surveyed companies, the Chemical Industrial Park Knapsack is the one that is least accessible for visitors; this may be one factor that explains the low number of visitors in this company park, and the plans of the company to offer a free shuttle bus service for visitors.

Benefits in terms of income and image for city and enterprise

First, let's consider the impact of industrial tourism on the image and income of companies. In Cologne, only the Magic Media Company reaches a small financial benefit out of its industrial tourism activities: the company charges visitors for the participation in the studio tour. The profit is, however, not the reason for MMC to carry out the tours. MMC is rather defensive when it comes to company tours – it was basically the attractiveness of the product 'TV' that has led to an overwhelming interest which more or less forced the company to open up. From what we can judge, the tours at MMC are rather considered a burden than a great benefit, and this is why the company at least tries to cover the financial costs for this burden. All of the other companies expect intangible benefits that are a lot higher than the cost of organising company tours, which is why – together with the ambition to attract a big number of people – they decided to offer the tours for free. For example, the industrial park Knapsack and the Bayer Group experience a lack of skilled labour, and expect that company tours can be a way to show themselves as an attractive employer. At Chemical Industrial Park Knapsack, industrial tourism is expected to create conditions that are conducive for the recruitment of high quality staff in general, also in areas where there is no personnel shortage. Another interesting example in this respect is the Rheinenergie AG: this company uses industrial tourism to get feedback from customers and to adjust their company policies accordingly; thus industrial tourism is a feedback mechanism of strategic value. Moreover, this company discovered industrial tourism as a tool to convince people that hold a negative opinion about the company, and this way improve the image of the company and increase customer loyalty. For the Bayer Group, the major aim of industrial tourism is to correct the image of the company – many people associate Bayer with chemical industry, which does not correspond to today's situation anymore. Industrial tourism is meant to show the full range of activities of the Bayer group and to make clear that Bayer is more than just 'Aspirin'. The case of the HGK is quite unique, as the company itself does not have many important interests in industrial tourism. This can basically be explained by the fact that this public company was urged by its owner – the municipality of Cologne – to open up.

So obviously, the city expects industrial tourism to have a certain impact on the image and income of Cologne. Especially in the case of HGK and the Rheinauhafen-project, a deep interest of the city is evident. Cologne is not only a place of thriving industries, it is also a city that suffers from ongoing structural change, a city that needs to attract new jobs and feels urged to further expand its service industry. By opening up this prestige project on the riverside to the public, Cologne tries to highlight certain advantages of the city which could create an image that is conducive and can attract jobs and service companies. This should help to achieve the aim of attracting 20,000 new jobs in that area. The Rheinauhafen project is exemplary for the situation in Cologne. As the city already possesses a thriving and diverse tourism industry, the development of industrial tourism as such has not been included in the city's strategy. However, in certain areas the promotion of industrial tourism visits has been recognised as a fruitful tool to improve the image of the city and to support existing strengths. This is why the city does pay attention to industrial tourism in the media sector. Many people are aware that Cologne is Germany's media capital: the media sector was one of the top ten things that people name when questioned for their associations with the city of Cologne (IHK Cologne, 2007). Company tours in this industry could contribute to further strengthening this image. On the other hand, too much focus on the topic 'industry' in the communication of the city is expected to be somewhat implausible, mainly because Cologne promotes itself with the slogan 'Cologne is a feeling'. Industry as such is considered to be too sober to fit with this message. Nevertheless, this doesn't keep Cologne from promoting industrial tourism in a more subtle way: company tours hold a quite central position in the city's event Expedition Colonia. However, the term 'industry' is avoided completely in communication to the public, although about 70 per cent of the Expedition Colonia-programme for adults, and about 30 per cent of the programme for children can be considered as industrial tourism according to our definition. These two programmes are mainly used to increase the inrush of visitors and the hotel occupancy rate during low demand periods. We have seen that the city of Cologne reaps a number of intangible benefits from industrial tourism, mostly related to the image of the city. On the contrary, the city does not receive any direct financial benefits from industrial tourism.

Concerning the question whether there are converging interests between the city and companies, the answer is rather ambiguous in the case of Cologne. In the few areas where the city of Cologne has an intense interest in industrial tourism, the private actors themselves have not (yet) fully recognised this interest (like in the case of the HGK) or are rather passive (in the case of the Magic Media Company). On the other hand, many companies were willing to participate in Expedition Colonia, which obviously offers benefits for both sides, public and private. Also in another area the benefits for both sides are clear: all public initiatives in Cologne that concern industrial tourism also involve companies from the surrounding Cologne region. We can observe that it is mostly big industrial attractions in the region that are included in the lists of the IHK and the CTB (like the Bayer Group in neighbouring Leverkusen or the Magic Media Company in the suburb of Hürth), whereas in Cologne itself also rather smaller institutions and companies are included (like the theatre). We could argue that especially industrial tourism for incoming delegations should be organised region-wide, rather than for the city only. Investors typically do not care for administrative boundaries, and it would be disadvantageous for a city not to include an important company that is located in another municipality nearby. Also for leisure visitors this is an interesting strategy that leads to a win-win situation: for tourists, this means an increase of the perceived amount of attractions, and hence an increased attractiveness of the city as a tourist destination. In return, the companies outside the city get a higher visibility and free promotion. Remarkably, Cologne does not view excursionists as a problem for the city – like other cities, a big share of industrial tourists might decide not to spend the night in the city in question. The city of Cologne is not at all concerned about the situation that many of their visitors are excursionists, due to the fact that many of these excursionists leave a lot of money in the city (IHK Cologne, 2007). What does this mean for industrial tourism? Cologne is known as a shopping city which makes it very likely that also industrial tourists that just pay a visit of one day will come to the centre of Cologne to dedicate some of their time to shopping. So also if industrial tourists don't spend the night in the city, they can be expected to be fruitful for the city in financial terms.

Organisational setup

Concerning the internal organisational setup of industrial tourism at the investigated companies, we made a number of interesting observations: the Bayer Group, the Rheinenergie AG and the Industrial Park Knapsack have integrated the organisation and planning of industrial tourism in their public relations or communications department. In the case of the Chemical Industrial Park Knapsack, InfraServ Knapsack is the owner of the premises and intends to organise industrial tourism on behalf of the companies on the premises. Bayer and Rheinenergie are probably the most professional and mature companies in organisational terms – both companies employ a couple of employees that exclusively deal with industrial tourism. In the case of Bayer this department is, however, only responsible for the arrangement of visits to the Bayer exhibition centre (the Baykomm), but not for the arrangements of company tours (which, anyhow, is not a top priority of the Bayer Group). Industrial tourism is rather weakly rooted at HGK and MMC. The studio tours at MMC are organised on the side by one project manager of MMC, additionally the company employs around 15 students (part-time employees) as guides. At HGK, tours of the Rheinauhafen-building site for professional visitors are organised by the real estate division of the company, whereas an external guide takes care of leisure visitors. Interestingly, leisure visitors have to pay for the visit, whereby only the guide makes a profit, while the company receives a symbolic contribution of €1.00. Thus, apart from the very professional Bayer Group and Rheinenergie AG, other companies might also learn from the interesting organisational model of the HGK, where only the responsibility for professional visitors has been retained within the organisation. The Bayer Group made the deliberate decision to put an emphasis on an exhibition centre, instead of company tours. There are various reasons for this: one reason is that more people can be admitted during a certain timeframe when compared to a factory visit, another reason is that the exhibition centre cannot cause any disturbance of the production process. According to our judgement, none of the surveyed companies has reached its capacity limit for receiving industrial tourists; the main limitations are a lack of personnel, or a lack of willingness to employ the personnel that is needed for the organisation of visits. Thus, a further inrush of visitors would not necessarily be disturbing for the production process. Only the Magic Media Company seems to be close to its capacity limit during certain very intensive weeks. These very pronounced seasonal fluctuations in demand can be explained mainly by the fact that MMC is mostly dependent on one single target group (school classes). Diversifying one's target groups can help to avoid heavy seasonal fluctuations in visitor demand. Hereby we can see that a passive behaviour in industrial tourism can lead to organisational difficulties. As MMC does not try to influence the composition of its target groups actively, they basically attract one blend of people with similar behaviour that leads to pronounced seasonal fluctuations.

The city of Cologne provides an organisational environment for industrial tourism that reflects the fact that industrial tourism is no prime initiative of the city. As we have explained before, the task of promoting industrial tourism is split among two organisations. While the Cologne Chamber of Commerce takes care of incoming professional delegations (most of which foreigners), the Cologne Tourist Board takes care of the promotion of company tours for leisure visitors and students. For both organisations industrial tourism is a rather insignificant side activity, which is also evidenced by the fact that these institutions do not possess a department that exclusively deals with industrial tourism. While this cannot be considered a best practice for industrial tourism, other cities can nevertheless learn from the Cologne model. For example, the Expedition Colonia – a public private partnership – does provide a good example for other cities and shows the way how industrial tourism can be used to tackle a lack of touristic demand in the off-season. It also shows in which way a public-interest platform can take over the arrangements of dates and booking from the companies themselves (also if this is only done temporarily during the period of the event).

References

Airliners.de 2006. *UPS Airlines.* [Online]. Available at: http://www.airliners.de/airports/sonderthemen/ups/2.php [accessed: 2008].

Bayer 2008. *Bayer AG.* [Online]. Available at: http://www.bayer.de/ [accessed: 2008].

Baykomm 2008. *Bayer Communication Centre.* [Online]. Available at: http://www.baykomm.bayer.com [accessed: 2008].

Bonn Region 2008. *Sightseeing.* [Online]. Available at: http://www.bonn-region.de/deutsch/sightseeing/voreifel-bruehl-koeln/tour-zum-koeln-bonn-airport.html [accessed: 2008].

Cologne Bonn Airport 2008. [Online]. Available at: http://www.airport-cgn.de/main.php?id=78&lang=1 [accessed: 2008].

Cologne Statistical Yearbook 2006. *Statistisches Jahrbuch Koeln 2006*, 85th ed. Cologne: Stadt Koeln. [Online]. Available at: http://www.stadt-koeln.de/imperia/md/content/pdfdateien/pdf/statistikjahrbuch/3.pdf [accessed: 2008].

Cologne Tourism 2007. *Meeting Point Cologne – the Congress and Event Area.*

Cologne Tourism 2008. *Sales Guide Greater Cologne 2008.*

Cologne Tourism 2008b. *Cologne Bonn Aiport – Flughafenbesichtigung fuer Jedermann.* [Online]. Available at: http://www.koeln.de/tourismus/koelntourismus/cms/artikel.php/24/513/artikel.html [accessed: 2008].

Expedition Colonia 2008. *Expedition Colonia – das Stadtentdeckungsfestival.*

Expedition Colonia 2008b. [Online]. Available at: http://www.expedition-colonia.de/ [accessed: 2008].

GAMESCom 2008. [Online]. Available at: http://www.gamescom.org/ [accessed: 2008].

HGK 2008. *Häfen und Güterverkehr Köln AG.* [Online]. Available at: http://www.hgk.de/ [accessed: 2008].

IHK Cologne 2006. *Was tut die IHK Köln im Tourismus?*

IHK Cologne 2007. *Shopping Tourismus in Köln: Potentiale nutzen, Chancen identifizieren!*, Cologne: Chamber of Commerce and Industry (IHK Cologne).

IHK Cologne 2008. *Tourismus und Gastronomie; Statistics.*

InfraServ Knapsack 2007. *Chemiepark Knapsack – Innovativer Chemie- und Industriestandort mit Zukunft.*

InfraServ Knapsack 2008. [Online]. Available at: http://www.InfraServ-knapsack.de/en [accessed: 2008].

LDB NRW 2007. *Structural data for the city of Cologne.* Cologne: Landesdatenbank Nordrhein Westfalen, Federal agency of statistics NRW. [Online]. Available at: http://www.lds.nrw.de/kommunalprofil/k05315.pdf [accessed: 2008].

Magic Cities 2008. *Magic Cities Germany.*

MeineStadt.de 2008. [Online]. Available at: http://www.meinestadt.de/koeln/statistik?Bereich=Wirtschaft+%26+Finanzen [accessed: 2008].

MMC 2008. Magic Media Company GmbH, [Online]. Available at: http://www.mmc.de/ [accessed: 2008].

Rheinenergie AG 2008. [Online]. Available at: http://www.rheinenergie.com/lang/en [accessed: 2008].

Rheinenergie AG 2008b. *Koelner Trinkwasser*, brochure on drinking water section of Rheinenergie AG.

Rheinenergie AG 2008c. *Fernwaerme im Fluss*, brochure on long-distance heating of Rheinenergie AG.

Schokoladenmuseum 2008. [Online]. Available at: http://www.schokoladenmuseum.de/ [accessed: 2008].

Stadt Köln 2008a. *Weltjugendtag.* [Online]. Available at: http://www.stadt-koeln.de/weltjugendtag/.

Stadt Köln 2008b. *Bevoelkerung* [Online]. Available at: http://www.stadt-koeln.de/zahlen/bevoelkerung/artikel/08955/index.html [accessed: 2008].

Stadt Köln 2008c. *Tourismus.* [Online]. Available at: http://www.stadt-koeln.de/zahlen/tourismus/artikel/04536/index.html [accessed: 2008].

Stadt Leverkusen 2008. [Online]. Available at: http://www.leverkusen.de/freizeit/tipps/sp_auto_591.cfm?men=stadtportrait&sub=sehenswertes [accessed: 2008].

UPS 2008. United Parcel Service Inc. [Online]. Available at: http://www.ups.com/europe/de/gerindex.html [accessed: 2008].

Willkommeninkoeln.de 2008. *Technical Visits*, [Online]. Available at: http://www.willkommeninkoeln.de/03tour/tour05-3e.htm [accessed: 2008].

Discussion partners

Mr. Dirk Binding, Head of Tourism and Hospitality, Cologne Chamber of Commerce and Industry.

Mr. Franz-Xaver Corneth, Head of project development (real estate), Häfen und Güterverkehr Köln AG.

Mr. Günther Geisler, member of staff of the communication department, Chemical Park Knapsack (InfraServ Knapsack).

Mr. Christoph Hanke, Deputy Director & Head of Foreign Trade Consulting, Cologne Chamber of Commerce and Industry.

Mr. Thomas Kuhlow, Head of communication, Chemical Park Knapsack (InfraServ Knapsack).

Ms. Nicole Müller, Project manager, Magic Media Company GmbH.

Mr. Klaus Odenthal, Head of marketing, Cologne Tourist Office.
Mr. Frank Straube, member of staff of the communication department, RheinEnergie AG.

Chapter 4
Industrial Tourism in the Pays de la Loire Region

1 Introduction

In this chapter we discuss the development of industrial tourism in the French Pays de la Loire region, one of the most active European regions in the field of 'economic discoveries' as the French prefer to call it. In contrast with other chapters we focus on the region rather than on one city. The reason is because one of the most interesting initiatives is taken on a regional level: the association 'Visit Our Companies in the Pays de la Loire Region' (*Visitez nos entreprises en la Pays de la Loire*), hereafter abbreviated as Visit Our Companies. This association is playing a key role in assisting companies with opening their doors, taking away barriers, improving quality and promoting industrial tourism in general. Interestingly, the companies that open their doors are active in different sectors, also in sectors that are not consumer-oriented (business-to-business). Nantes is the largest city in this region, but our analysis mainly focuses on companies in two neighbouring towns: the city of Angers (which hosted the first European conference on company visits in 2006, and invented the Made in Angers event in 2000) and St Nazaire (a port city with two major companies that opened their doors: Airbus and STX Europe).

The structure of this chapter is as follows. Section two presents some general information on the Pays de la Loire region and the three cities that we visited, followed by a tourism profile. After that we concentrate on the organisational framework for industrial tourism, introducing several organisations that are active in this field. The fourth section describes and analyses the industrial tourism activities of individual firms that we have selected in cooperation with Visit Our Companies. The final section confronts the observations with our frame of analysis that we developed in Chapter 1, drawing conclusions on the conditions under which industrial tourism has been developed.

2 General profile of the Pays de la Loire region

Pays de la Loire ('lands of the Loire river') is one of the 26 regions of France[1] and consists of five departments: Loire-Atlantic (including the cities of Nantes and St Nazaire), Maine-et-Loire (including the city of Angers), Mayenne, Sarthe

1 Four regions are overseas territories.

and Vendée. It is one of the regions created in the late 20th century to serve as a zone of influence for its capital, Nantes, one of a handful so-called 'balancing metropolises' (métropoles d'équilibre). In 2005, the region had a population of 3.4 million (the fifth region of France), of which 1.2 million lived in the Loire-Atlantic department.

With 281,000 inhabitants, Nantes is the sixth largest of France (after Paris, Marseille, Lyons, Toulouse and Nice) (2005 data); its metropolitan area counts approximately 545,000 inhabitants (1999 data), resulting in the eighth position (after the five cities mentioned above and Bordeaux and Lille)[2]. Nantes is the capital of both the department Loire-Atlantic and the Pays de la Loire region. The metropolitan region of Nantes has developed into an attractive alternative for Paris. More and more companies from the Paris region decide to relocate to Greater Nantes, a phenomenon known as the 'West Coast Effect' (*l'Effet Côte Ouest*). In France, the Nantes region has a very positive image, but abroad, the city has no clear image.

Since our analysis is not restricted to Nantes, but also comprehends the development of industrial tourism in Angers and St Nazaire, some general information on these cities has to be provided too:

- Angers (150,000 inhabitants) is the capital of the department Maine-et-Loire. Its nickname La Ville Noire (the black city) refers to the slate on the roofs of many buildings that used to be quarried in the nearby mines. Today the city has a diversified economy and a university within its borders. Among the sectors that are well represented in the regional economic structure are the liqueur industry (Cointreau being distilled here), lorry manufacturing, computer manufacturing and research in horticulture and biotechnologies.
- St Nazaire (68,000 inhabitants) was founded in the 19th century with the intentions to develop a port for the city of Nantes. In the Second World War the city was bombed, explaining the lack of an attractive town centre. St Nazaire is clearly an industrial town with a strong presence of not only the port, but also of some large companies such as shipbuilder STX Europe and aeroplane builder Airbus. Some former port areas of St Nazaire are currently redeveloped into an attractive centre combining residential, commercial and leisure functions (Ville Port, 2008).

The tourism profile of the Loire Region

Statistics on tourism in the region are provided by the Regional Observatory for Tourism. In 2004, hotels in Pays de la Loire registered 3.4 million arrivals and 5.7 million bed nights, of which 55 per cent in the summer months (May–September). International visitors spent 769,924 nights all together (a share of 13 per cent). Great

---

2 Source: www.citypopulation.de and Institut National de Statistique et des Études Économiques.

Britain and Ireland are by far the most important countries of origin, consuming almost one third of all bed nights. Other important countries of origin are Belgium, Germany, Italy, Spain and the Netherlands. In the same year, camping sites in Pays de la Loire counted 1.4 million tourists spending 9.5 million nights in total, of which 83 per cent in July and August. With 2,192,266 bed nights, international visitors take a share of 23 per cent in this number. Again Great Britain and Ireland are the main foreign suppliers of tourists (48.7 per cent), with the Netherlands as runner up (33.2 per cent) (Observatoire Régional du Tourisme des Pays de la Loire, 2005). With a share of 5.9 per cent in all bed nights registered in France, Pays de la Loire cannot be seen as one of France's top tourist regions. In absolute numbers, Rhone-Alpes, Provence-Alpes-Côte d'Azur and Ile-de-France (Paris) are the top tourist destinations, with a relatively high share of foreigners (non-residents). With between 20 and 30 bed nights per capita, Pays de la Loire is also less 'touristic' than the neighbouring regions Brittany (to the northwest), Lower Normandy (to the north) and Poitou-Charentes (to the south).

Pays de la Loire present many great monuments as the castles of Angers, Laval, Mayenne and the Nantes 'Château des Ducs de Bretagne', the Royal Fontevraud Abbey (the widest monastic ensemble in Europe), the old city of Le Mans but also many Natural Parks (such as the Brière and the Marsh of Poitou). In 2004, 500 touristic sites registered more than 9 million entries. Monuments and museums represent respectively 25 and 21 per cent of all visits. Company visits take a share of 11 per cent, which implies an absolute number of almost 1 million visitors a year. The most visited industrial tourism attraction is Terre de Sel in the town of Guérande, with 38,322 visitors in the year 2004. With more than 1.2 million visitors in 2007 Le Grand Parc et La Cinéscénie du Puy du Fou is the most popular attraction; in the three cities we visited (Nantes, Angers and St Nazaire), the main attractions are the Fine Arts Museum of Nantes (95,087 visitors in 2007), the Angers Castle (152,733), the Fine Arts Museum of Angers (74,163) and the Ocean Liner Museum Escal'Atlantic in St Nazaire (219,887) (Observatoire Régional du Tourisme des Pays de la Loire, 2007).

A recent study among visitors of four major sites in the Loire region – based on more than 1,500 interviews with tourists – gives some insight into the combinations tourists make on one day. It appeared that about 2 per cent of the visitors made a combination of visiting one of the four major sites with a company visit (visits to wine producers not included; they represent another 2 per cent) on the same day. Looking at three consecutive days, approximately 3 per cent of the visitors made such a combination (SEM Régionale des Pays de la Loire, 2007).

3 The organisational framework for industrial tourism in the Loire Region

In this section we introduce three organisations that are particularly relevant for the organisational framework for industrial tourism in the Loire Region. First we discuss the event Made in Angers, a joint initiative of several actors in the Angers region. Second we analyse the activities of Visit Our Companies, an initiative on the level of Pays de la Loire. The third sub-section is about Escal' Atlantic, which is the tourism organisation for St Nazaire.

Made in Angers

The first initiative in the field of industrial tourism in Pays de la Loire was taken in 2000 when the local Chamber of Commerce and the local and metropolitan tourist offices invented the annual event 'Made in Angers' which takes place in February/March. This event started with eight participating companies and four craftsmen that opened their doors to the public for one week, and developed into a four-week event with more than 100 participating companies and 50 participating craftsmen from the Angers region, drawing about 17,000 visitors a year. Students and pupils have free entrance, while adults pay between €3 and €5 for one company visit. There is also a discount ticket that gives access to four sites for the price of three. Among the participants are INRA (centre for agronomic research) and Carré Cointreau (liqueur); both to be introduced in the next section. In total, 15 sectors are represented including the metal industry, agriculture, services, media and communication, architecture, the environmental industry, research, education et cetera.

During two weeks of the event, only individual visitors can participate in company tours (visites grand public), whereas two other weeks are reserved for group visits (students and school classes). Every year, the event has some new innovative elements: in 2008, every weekend was dedicated to a different topic (local crafts, gastronomy, technical museums). Moreover, the 2008 edition had a special focus on particular professions: at the end of each guided tour, every company gave a presentation about one specific job that is practiced within the firm in question. Additionally, 24 companies gave a presentation about their policies on sustainable development.

The main occasion for the event was to strengthen the relation between local business and the local population, in view of the growing gap between companies and citizens: many French people hold a negative view on the corporate world. Another motive to set up Made in Angers has been to strengthen the relation between educational institutions and business, and more in particular, to change the attitude of schools and universities: in France many educational institutions still tend to focus on providing knowledge, while they pay less attention to preparing children and students for a job. One of the associated partners in Made in Angers is the Angers University (ESTHUA), which even developed a professional bachelor's programme in 'economic discoveries' (Licence Professionelle Découverte Economique[3]) in cooperation with local and regional Chambers of Commerce and

3 A professional bachelor (licence professionnelle) is more specialised than a regular bachelor, and is particularly relevant for students want to start working right after they receive their bachelor's degree. More informationa about the course can be found on http://www.univ-angers.fr/formation.asp?ID=ELPDECO&langue=1.

Tourism Offices and the association Visit Our Companies, to be introduced in the next sub-section. The programme includes courses in communication (foreign languages), quality management, industrial heritage and industrial tourism, and provides knowledge on the regional economy. Moreover, students are expected to do a traineeship at one of the active companies, or the associated organisations (e.g., Made in Angers or Visit Our Companies).

Another result of the cooperation between the Angers Tourism Office, the Angers Chamber of Commerce (as well as the national Chamber of Commerce) and the University of Angers is an international conference on company visits, which took place in Angers in March 2006. With this conference the initiators wanted to create a platform for exchanging experiences in the realm of industrial tourism. It has been decided to organise such a conference every two years; the 2008 edition took place in Toledo (Spain).

Visit Our Companies

In 2001, the decision was taken to create the association Visit Our Companies in Pays de la Loire (*Visitez nos entreprises en Pays de la Loire*) in response to a study carried out by the Regional Observatory of Tourism in 1999. This study made clear that several companies in the region were interested in opening their doors to the public, while other companies that were already open expressed their wish to share experiences in order to improve their supply. Moreover, the initiators expected that the demand for 'economic discoveries' would grow: people are increasingly interested in their roots, in the working life of their fathers and forefathers, and in the 'economic identity' of the region they live in or they visit.

The association is a joint initiative of the regional committee of tourism, the regional Chamber of Commerce and Industry (CCI) – the local chambers and an initial core group of 24 companies. The professional organisation is (financially) supported by the region and the State. The regional committee of tourism and the chamber of commerce both finance half of the budget, the other half has to be financed by the member companies, by donations and through commercial revenues. For the region, the development of industrial tourism fits in a broader framework of developing 'economic discoveries' which also comprehends the development of science tourism and industrial heritage tourism.

Visit Our Companies has only one permanent staff member: the managing director of the association. The organisation's board of managers (one president and six vice-presidents) consists of representatives from member companies while the general secretary is provided by the Chamber of Commerce and Industry (they are all not employed by the association). The first president of the association was a CEO of Airbus, while the current president (2008) is a CEO of Electricité de France, an energy company.

Table 4.1 Members of Visit Our Companies (with selected companies in bold)

Category	#	What companies?
Recycling and environmental industries	5	Valorena (garbage treatment and recycling), Arc-en-siel (garbage management), EDF New Energy (wind turbines), Lyonnais des Eaux (water purification), Salamandre Sivert de l'Est Anjou (energy production and distribution, using garbage)
Agrofood industries	4	Altadis (cigarettes), Bahier 'Au Pays Des Rillettes' (meat), La Mie Caline (industrial bakery), Le Lactopole (dairy)
Wood, paper, textile and leather industries	7	Fonlupt (clothing), Goubault (printing), Les Ateliers Le Mauf (silk screen), Loiseau Menuiserie (carpenter), Pedi Girl (shoes), Sarl Petiot (football and billiard tables), Sarl Vincendeau (furniture renovation)
Other industrial activities	7	**Airbus St Nazaire** (aeroplanes), **STX Europe** (shipbuilding), **The Autonomous Port of Nantes-St Nazaire** and four other producers of respectively zirconium (used for nuclear reactors), granulates, energy and medical equipment.
Artisanal food industry	13	**Terre de Sel** (salt) and 12 other producers of jam, cheese, chocolate, mushrooms, snails, biscuits, syrup and flour mill.
Distilleries and breweries	14	**Carre Cointreau** (liqueur) and 13 other producers of wine, beer and liqueur.
Handicraft	7	
Services	1	CCO Nantes (permanent exhibition about the regional economy)
Agriculture	4	**Terres des Sciences** (fruit and trees), and three other agricultural companies.
Others	4	

Source: www.visiteznosentreprises.com

The main objectives of the association are to promote the knowledge base of regional companies, to develop industrial tourism in the region, to stimulate exchange between the different members of the association, to professionalise the activities and to measure visitor numbers (Visit Our Companies, 2008). In 2008, the association counted 75 members. They all contribute to the continuity of the association by paying a fee depending on the number of employees: companies with more than 50 employees pay €500 per year, whereas smaller companies pay €230. The table above gives an idea of what kind of companies are member of the association. The food sector is well represented, with about 50 per cent of the memberships.

Looking at its objectives, the association's main target group are not visitors, but companies. Assuming that the demand for company visits will grow anyway, the organisation's focus is on improving the supply of company tours, quantitatively and qualitatively. Visit Our Companies does not arrange visits and does not issue tickets: the association tries to bring companies in a position that they themselves are able to organise tours. Through its website (www.visiteznosentreprises.com) the association presents itself as an intermediary for interested parties, referring them to companies and travel agencies that do organise the tours. Another, more international, portal for (company) tours is the website of the regional committee of tourism (www.enpaysdelaloire.com) which provides information in seven languages (French, English, German, Dutch, Spanish, Chinese and Japanese)[4].

Another means of communication is a practical guide updated every year of which 50,000 copies are distributed via the Regional Tourism Committee, the Chamber of Commerce, the network of consulates, tourism offices, national and international exhibitions, tour operators, and schools. In its role as intermediary, the association also organises press meetings (on average two a year) and special tours to one of the member companies (so-called Eductours) for tour operators and travel agencies. Several meetings with regional and national media made it possible to inform a new audience about industrial tourism. Only in 2005, more than 60 media articles appeared concerning meetings organised by the association, or reports on the member companies (Charuel, Leblanc and Brin, 2006). Furthermore, the association is responsible for organising the national open door days (*Journées Nationales de Portes Ouvertes en enterprises*) in the Loire region. Through this event Visit Our Companies not only reaches a larger audience (promoting the industrial tourism product), but it also helps to convince companies to open their doors, at least for once a year, and maybe to become member of the association later on. In 2005, more than 60 non-members were participating in the manifestation, opening their doors for the first time to the public (Charuel, Leblanc and Brin, 2006).

One of the main tasks of Visit Our Companies is quality control. The organisation played a pilot role in the introduction of a national quality programme Destination Enterprises (Destination Companies). All members are assessed by a quality commission, which consists of one representative on behalf of the Regional Tourism Committee, one from the Chamber of Commerce and Industry (regional or local), one representing the member companies and the managing director of the association. The quality commission looks at several conditions, some of which are obligatory for membership while others are 'preferable' (Visit Our Companies, 2008):

- The association can only be joined by active companies that allow visits to their production site or a 'site vitrine' (display site).
- The objective of the site should not be strictly commercial: the company should present its technology, knowhow and knowledge in an active and lively way.

4 In 2009 a new Web 2.0 based website has been launched enabling all actors in the tourist industry to promote their offer, and giving customers the opportunity to give their opinion on the supply of tourism.

- Sufficient parking facilities are obligatory, while a reception, a room for presentations, toilets and an attractive immediate environment (including signing) are recommended.
- The presence of at least one 'reference person' who understands how to receive guests and has sufficient knowledge about the company (obligatory), and who speaks one or more foreign languages (at least English) (preferred). Visits are preferably supported by documentation, video, etc.
- Companies have to provide a reference paper and they have to guarantee the safety of the visitors (also by taking out insurance).
- Companies have to accept individuals in groups, and they should communicate their opening days and hours, as well as the tariffs for visiting the company; visits should be possible at least six months of the year, except for seasonal production. Preferably a site is open at least three months between May and September and at least four days a week (Monday to Friday), or alternatively three ways a week and during the weekend.
- The companies also express their commitment to the association by participating in promotional efforts, and by supplying visitor figures regularly.

To further develop industrial tourism, the regional Chamber of Commerce and the managing director take care of recruiting new members. To professionalise the offer of member companies, the association organises several meetings such as workshops (free of charge for members) that deal with a specific theme selected by the companies themselves (e.g., how to deal with foreign visitors or how to get funding for developing a full day programme). Two general assemblies a year take place at a site of a member company, enabling other firms to learn from this particular company. Every two months a newsletter is sent to the members and partner organisations.

Another task of the association is to keep track of visitor numbers. Members are obliged to send visitor numbers, at least once every year. In practice, however, the collection of data appears to be problematic, as many companies need to be reminded that they have to send the latest visitor figures. There are no clear incentives or penalties, although it is an obligation in the quality chart. Anyway, all sites of industrial tourism have been integrated in the database of tourist sites in the region (Cyvel Sats[5]), developed by the Regional Observatory of Tourism.

Escal'Atlantic

Another organisation that is active in industrial tourism is Escal'Atlantic, the tourism organisation for St Nazaire, of which 70 per cent is owned by public shareholders (e.g., the regional Chamber of Commerce and Industry), and 30 per cent by private stakeholders (mostly banks). The foundation of Escal'Atlantic in the late 1990s can

5 Cyber Voyage en Ligne.

be seen as part of a comprehensive strategy to make the city more attractive, not only in terms of tourism but also as a place to live and to work. Also the waterside redevelopment project 'Ville Port' fits in this strategy. The organisation employs about 50 people, including five professional guides. Escal'Atlantic owns and operates the Ocean Liner Museum (which was opened in 2000) and the Submarine Espadon (which opened by the end of the 1980s), two important tourist attractions in St Nazaire. In addition, the organisation arranges tours to three companies in the Angers region: the Autonomous Port of Nantes-St Nazaire, STX Europe and Airbus.

In total, the attractions of Escal'Atlantic receive about 200,000 visitors, in which the three company visits take a share of about one third. The majority of the visitors are holidaymakers from the nearby beach resorts; city-trippers whose primary purpose is a visit to the city of Nantes also come to St Nazaire, but not in large numbers. For all three companies, Escal'Atlantic provides all equipment that is needed for the organisation of guided tours, but also invests in visitor facilities such as footbridge in one of the assembly halls at Airbus. Escal'Atlantic has signed multi-year contracts with the three companies in which they got the exclusive right to organise company tours for leisure visitors.

4 Industrial tourism products of Pays de la Loire

In this section we discuss some examples of industrial tourism products in Pays de la Loire: the Autonomous Port of Nantes-St Nazaire, STX Europe, Airbus, Cointreau, Les Salines de Guèrande (Terre de Sel) and INRA Angers-Nantes (Terre des Sciences). They are all members of Visit Our Companies, and represent several – though not all – sectors that open their doors. The companies have been selected in consultation with Visit Our Companies.

The Autonomous Port of Nantes-St Nazaire

The Autonomous Port of Nantes-St Nazaire is a public enterprise corporation, owned by the State, which is in control of around 20,000 acres along the Loire River estuary. The Port of Nantes-St Nazaire is the fourth largest autonomous port authority of France, behind Marseille (95.5 million tonnes), Le Havre (69 million tonnes), Dunkirk (50 million tonnes), and ahead of Rouen (21.9 million tonnes) and Bordeaux (8.4 million tonnes). At the end of 2007, the Port Authority employed 710 people (Autonomous Port of Nantes-St Nazaire, 2007; 2008).

The main sites are located in the municipalities of Donges and Montoire de Bretagne, located 60 km downstream from Nantes. Donges is the main site for the handling of oil, while Montoire de Bretagnes accommodates specialist terminals for handling RoRo (Roll-on Roll-off), containers, liquified natural gas, agrifood and coal. In the near surroundings of Nantes there are terminals for grain, sand and forest products in Chevirė and Roche Maurice. A cruise liner terminal is the only port activity that can still be found on the Nantes Island (Île de Nantes), before the 1970s the main location of the port. Finally, St Nazaire is specialised in shipbuilding and ship repair, and accommodates several specialised terminals (notably food and heavy cargo).

Already in the early 1990s, the Port started to recognise the importance of industrial tourism. Since 1988 the number of visitors has increased from 1,800 to about 10,000; most of them are French, only an estimated 10 per cent is foreigner. Visitor statistics show that the group between 25 and 40 is clearly underrepresented. Leisure visitors account for the biggest part of the indicated 10,000 visitors; there are only around 10 to 15 professional delegations a year, with an average of around 5 people per group. Many leisure tourists, however, do have a strong relation with the port: e.g.,farmers are interested in the agrifood terminal where the soya is stored they buy for their cattle. Hence, in these cases leisure motives are combined with business motives. Many of the port's visitors come from the Pays de la Loire Region itself, but also other regions account for a strong inrush of visitors, notably regions that constitute the hinterland of the port and regions that also have a seaport (e.g., La Rochelle, Bordeaux).

The main motive for the Port to organise tours is the need to communicate the added value (generating 26,000 jobs in the region) and to improve the image of the Port and port-related companies as employer (attracting future employees). The Port Authority plays a crucial role in convincing companies in the port to open their doors, the need for transparency in view of image and reputation being one of the arguments used. The Autonomous Port of St Nazaire is an active member of Visit Our Companies. In 2005, the previously troublesome relation between the port and the city of Nantes improved considerably, mainly because both parties found each other in the redevelopment of the Nantes Island, which is to be developed into an extension of the city centre (Nantes Island, 2008).

On its website the Port refers to three tours that can be booked: a tour of the facilities in Nantes, Chevirê and Roche Maurice (with a length of 90 minutes), a tour of the facilities at St Nazaire, Montoir de Bretagne and Donges (with a length of 150 minutes), and a tour of all port facilities (which takes one full day in view of the distance between the facilities). All tours are by coach. Tours can be booked via the Communications Department, at least two weeks in advance. The organisation of the tours, however, has been outsourced to other actors. Some years ago, the Authority came to the conclusion that its capacity was insufficient to deal with the increasing number of requests for visitors. Visits to the sites near St Nazaire, Donges and Montoir de Bretagne are organised by Escal'Atlantic. These tours are not only in French, but also in English for groups of at least 20 persons. An interesting product is the Allovist card which provides a guided tour to the harbour using the mobile phone. Since May 2008, the tourist office of Nantes takes care of the visits to port sites in and near Nantes. The Communications Department plays a more strategic role, making sure that both organisations represent the Port in an appropriate way. Moreover the Department – in cooperation with Visit

Our Companies – assists companies in the port with professionalising corporate communication and company visits.

STX Europe

In June 2006, the Shipyard in St Nazaire, formerly known as 'Chantiers de l'Atlantic', became part of Aker Yards, an international shipbuilding group with headquarters in Norway and sites in Brazil, Finland, France, Germany, Romania, Ukraine and Vietnam. The group changed its name into STX Europe in 2008. In total the company employs 16,000 people, of whom 3,000 work in St Nazaire. The St Nazaire location, with a history that goes back to 1840, is specialised in the construction of cruise ships and big ferry ships; with a total surface of 108 hectares it is the second biggest location of the company, after the Turku location in Finland. One of the ships that were constructed here, is the Queen Mary II, the largest ocean liner of the world[6]. St Nazaire is one of the few locations in the world where cruise ships are constructed[7]. Only a handful of shipping lines can afford to buy these cruise ships, worth about €500 million. As the cruise industry is the fastest growing segment of the travel industry – with 2,100 per cent growth between 1970 and 2007! (MarketNewZealand.com, 2008) – the STX Europe workshops are operational 24 hours a day, 7 days a week.

In the early 1990s the St Nazaire shipyard started to open its doors for visitors. The main motives were: 1) to become more attractive as employer (improving the image of the shipbuilding industry, notably among higher-educated people); 2) to find new subcontractors (to keep prices low and to fulfil the growing demand); and 3) to secure good relations with the municipality (as the company needs the City for permissions). Within the company, the human resources and communications department is mainly responsible for organising company visits for professional visitors (technical visits), free of charge. The communications section, which is part of this department, invites journalists, and – in cooperation with the purchasing department – sub-contractors. The tailor-made tours for potential subcontractors are clearly business-oriented. The human resources section organises company tours for students from technical schools as part of their recruitment strategy. The human resources and communications department also takes care of communication towards the staff, explaining them the importance of company visits.

The organisation of tours for other groups with an educational motive and tourists with a leisure motive has been outsourced to Escal'Atlantic, notably because the company has not sufficient human resources itself. In some cases,

6 Ocean liners are different from cruise ships: an ocean liner is a ship designed to transport people from one seaport to another along regular long-distance maritime routes according to a schedule. A cruise ship or cruise liner is a passenger ship used for pleasure voyages, where the voyage itself and the ship's amenities are part of the experience.

7 There are actually only four countries in the world, all of them in Europe, that produce cruise ships: France, Finland, Germany and Italy.

even professional visitors are redirected to Escal'Atlantic. Visitors pay an entrance fee of €7; a part of this fee is transferred to STX Europe as compensation, resulting in an annual income for STX Europe of about €4,000. The co-operation with Escal'Atlantic is very close: their professional guides are constantly informed about changes at STX Europe, and they are also given the possibility to meet specialists in certain areas, in order to upgrade their knowledge about the company. In this way STX Europe still has considerable influence on the content of the tours. Over time the two organisations have established a level of mutual trust, which is also reflected in the observation that professional guides express their passion for the shipyard.

In a 'normal' year STX Europe receives around 30,000 leisure visitors, a number which has been rather stable over the past few years (only counting the visitors that come through Escal'Atlantic). Additionally, the company receives around 2,500 professional groups a year. The demand from visitors depends on the ship under construction. In 2003 the construction of the famous Queen Mary 2 led to an inrush of 150,000 visitors, resulting in a dramatic accident: a footbridge collapsed causing the death of 15 persons. The company experiences extreme peaks in visitor demand: many people request a visit just before a big ship is finished off. Company tours are offered seven days a week, only during the day. Visitors that come in through Escal'Atlantic receive a standard tour of 90 minutes. They arrive by bus and get to see most of the area by bus. Only one production hall and the dry dock can be visited by foot. In the production hall an elevated walkway has been installed allowing the visitors to witness the production process safely, and without disturbing the production. The dry dock, where the ships are put together, already had such a walkway before visitors came in; it only needed some small adjustments. The professional visitors receive a tailor-made programme that usually lasts around half a day and includes a presentation by a specialist in the interest area of the group. Only small professional groups can, in some cases, get the opportunity to walk inside the ships that are under construction. In view of safety and security it is not allowed to take pictures (to avoid espionage), and leisure visitors have to be registered at least two days before the visit (for professional visitors, this has to be done one day in advance). STX Europe does not handle a maximum size for visitor groups: large groups are simply split up into several groups, each with their own guide. Apart from the two elevated walkways, the firm has not invested in any other visitor facilities. There are no toilets available for visitors on the site, but the establishment of Escal'Atlantic is only a 5-minute drive from STX Europe and possesses all the necessary facilities. All leisure visitors start the tour on this location, making the lack of visitor facilities at the STX Europe site less problematic.

Of all locations of STX Europe worldwide, the St Nazaire location is the one that receives most visitors. It was already involved in industrial tourism before it entered the STX Europe Group. Being astonished by the success of the visits in France, the STX Europe headquarters have decided to professionalise visitor services in other locations as well, also in response to the wish of STX Europe' customers who consider industrial tourism as pre-sales promotion.

Airbus

The development of Airbus dates back to the late 1960s when politicians from France, Germany and Great Britain started to discuss the possibilities of developing passenger aircrafts in a consortium. This resulted in the development of the A300. Since then Airbus developed into one of the two leading manufacturers of aircrafts in the world[8], holding a share of about 50 per cent in the global market for the construction of passenger aircrafts. It is a 100 per cent subsidiary of the European Aeronautic Defence and Space Company (EADS), which is quoted on several stock exchanges in Europe. EADS, which employs about 116,000 people at more than 70 production sites worldwide, is the result of a merger in 2000 between DaimlerChrysler Aerospace AG (Germany), Aerospatiale Matra (France) and CASA (Spain). Airbus is one of the five divisions of EADS: it employs 57,000 people (Airbus, 2008). In France, the company has locations in four cities, of which two are located in the Pays de la Loire region: Toulouse (headquarters), Méaulte (near Paris), St Nazaire and Nantes. The two production sites in St Nazaire cover 50 hectares and provide employment to about 2,300 people. Both plants are specialised in the assembly of the front and central part of aeroplanes (the forward and central fuselage), and the testing of equipment. Upon completion, the forward and central fuselages are sent by ship or by cargo aeroplane to Toulouse or Hamburg (Germany), for the final assembly. The site in Nantes, with a workforce of 2,000 people, produces the centre wing box, the structural heart of the aircraft.

Airbus St Nazaire has a special department for the arrangement of professional and education visits. This department, which employs four professional guides, takes care of visits to the two production sites in St Nazaire. There are two standard tours: the long one takes two hours and starts at the Airbus visitor centre (including a small museum and models of aeroplanes), and then continues at the assembly hall and the equipment test hall. This tour costs €200 per group, with a maximum size of 25 persons (for bigger groups, more guides are provided, resulting in higher prices). A shorter version costs €150 per group, takes 90 minutes and follows the same route, except for the equipment test hall. The fees apply to companies without a direct business relation with Airbus: schools, customers and military personnel can usually participate for free.

The organisation of visits for tourists has been outsourced to Escal'Atlantic (in contrast with Toulouse where the company itself takes care of receiving tourists). Every year the local tourism organisation brings around 25,000 visitors to the Airbus plants in St Nazaire, often in combination with a visit to STX Europe. Tours take approximately 90 to 120 minutes. Leisure visitors come in by bus (they

8 The other one is Boeing.

get some general information about the company on their way to the plant), and get to see the assembly hall and the equipment test hall (the visitor centre is not included in the tour). The price of an admission ticket varies between €6 and €12, depending on the season, age of the visitor and discounts for groups and families. Escal'Atlantic can bring in visitors anytime during the agreed timeframe without prior announcement. Only on some special occasions – for example the visit of a government delegation – the access to the premises is denied for leisure visitors. Tours are in French, but English-speaking tours can be arranged for parties of at least 20 visitors.

The main motives for Airbus to be involved in industrial tourism are the need to attract qualified labour and the need to improve the image of the company, and more in general, the manufacturing industry. The firm wants to be seen as a good employer, but also as a good citizen that respects its environmental responsibilities. Moreover company visits are in the interest of the firm's customers: visits to Airbus can help to take away fear of flying (e.g., through the provision of information about quality control). Each year the guides who work for Escal'Atlantic get an update on the company's activities and its communication strategy. Every three months both organisations come together to discuss the visits. In January 2008, the guides were instructed to pay more attention to the need for human resources and the company's initiatives in the field of eco-efficiency. In response to the growing need for qualified labour, Airbus started in February 2008 to approach schools more proactively, inviting teachers and school classes to visit the plants in the Pays de la Loire region. As a consequence the share of visitors with an educational motive has increased from about one third in 2007, to about 50 per cent in April 2008. Another strategic initiative is *Rendez-vous d'Airbus* (Meet Airbus). Every month Airbus and Escal'Atlantic organise a thematic presentation and discussion (arranged by Airbus St Nazaire), followed by a guided visit to the site, with a total duration of 2 hours and 30 minutes. The prices of admission tickets are similar to the normal tours (€12 for an adult during the high season). These special tours take place on either Wednesday or Sunday. Sunday has been selected to make the tour more attractive (the plants are not operational during weekends) in view of declining visitor numbers. Airbus would prefer to have more tours during the weekends to reduce the risk of production delays and security risks. Escal'Atlantic, however, only wants to organise tours during the week (apart from the special Rendez-vous tours) because it is more attractive for visitors. Predefined routes for visitors on foot have already been introduced in the production halls, and have proven to be a successful tool to avoid production delays and ensure the safety of visitors. The risk problem will be further tackled by the introduction of an indicated route for coaches.

Cointreau

The Cointreau distillery in Angers was founded in the year 1849 and became famous with the production of its orange liqueur. Today, the distillery is owned by the Remy-Cointreau Group, the result of a merger in 1991, and it is still the only place in the world where Cointreau is produced. This factory employs around 200 people, of whom 60 are working in the production. Communication campaigns are developed in Paris where the group has its marketing department (25 people). Apart from production, the Angers site is also responsible for the distribution of all brands.

Already in the early 1920s Cointreau started to offer company tours; not by professional guides but simply by employees of the company who also had another job to do. Soon the factory became one of the biggest tourist attractions in Angers. Already in the 1930s Cointreau and the Angers Castle – the other major attraction – agreed to cooperate by redirecting visitors. In the late 1960s the company decided to improve its company visits, notably by investing in visitor facilities. When a new building was constructed in 1972 to replace the old factory, the reception of visitors was taken into account: the new building included a bar, a shop and an elevated walkway for visitors, which allows observing the production process in the bottling plant. This walkway not only has the function of an observation platform, it is also used as a corridor by the employees of Cointreau, and hosts a lot of historic photos and artefacts that are related to the company.

In 1999 (150 years after the foundation) the company decided not only to open an exhibition room about the history of the company (Cointreau museum), but also to give industrial tourism a more prominent position in the organisation, setting up an organisational unit that deals exclusively with company visits. This department is part of the marketing and communications division; between 1999 and 2008 the number of full-time jobs in this department doubled from two to four. These employees are responsible for the strategic alignment of the visits, and they also act as tour guides for professional and VIP visitors. The guided tours for leisure visitors are done by part-time employees (mostly students). These guides have to undergo a training that lasts two weeks, before they are allowed to guide visitors around the plant. Experience has shown that most guides look for another occupation after one year, which is basically attributed to the fact that the job is very monotonous (the guides only provide one kind of standard tour).

The Cointreau factory is open for public seven days a week from 11 a.m. till 6 p.m. Between November and April, however, there are no tours on Sunday and Monday, simply because there are few tourists in the region (low season). Although tours are less attractive during the weekends (because machines stand still), the number of visitor requests is high. Apart from the regular opening times, visitors can also visit Cointreau during the event "Made in Angers". Of all companies that participate in Made in Angers, Cointreau is the one that receives most visitors: the company is also generally considered to be one of the best examples for other companies that (want to) participate in this event. Interestingly, the entrance fee during the Made in Angers event is only €5, which is cheaper than the fee charged for a regular visit. According to the judgement of Cointreau, a quite different audience is attracted during the Made in Angers event, which manages to draw a blend of visitors that reflect the composition of the population more accurately than the regular visitors.

The standard tour to the Cointreau distillery takes between 75 and 90 minutes; one and a half hours are considered to be the maximum amount of time a tour should last. Experience has shown that visitors get bored and simply don't pay attention anymore if a tour takes too much time. At Cointreau, every tour starts with a ten-minute video about the company, which is shown in a building dedicated to visitors only (Carré Cointreau), located next to the plant. Afterwards, visitors head to the company museum (located in the plant, also part of Carré Cointreau) and can subsequently observe the production process – from the selection of ingredients to the bottling plant. Cointreau doesn't show all steps of production to visitors, but focuses on the most important ones. For safety reasons, the bottling plant can only be seen from the elevated walkway that runs along this facility. On some occasions visitors can also talk briefly to production workers.

In some cases, the order of the visit (film – museum – production) is changed, in order to make sure that visitors see the production line in operation. At the end of the programme, the visitors return to the visitor centre where the company-owned bar is located. There they can mix their own cocktail. After this, Cointreau offers the possibility to perform a breath-test and to check one's alcohol level. Visitors are also asked to fill in a feedback-form in order to evaluate the tour. Cointreau extracts additional information from the visitors by the means of this form – for example on how they got to know about the company tour (experience has shown that most people became aware of the tours through the internet, or through word-of-mouth recommendation)

Company visits only account for around 2 per cent of the overall promotional budget of the company. By collecting entrance fees, the department of visitor service can actually cover its costs completely (although the company visits don't produce any profit). In 2007, the decision was made to increase the fee considerably from €5.50 to €9.50. In return visitors get more quality, which is reflected in the reduction of the maximum group size from 60 to 25, but also in the complete refurbishment of the Cointreau bar.

The main reason for Cointreau to be involved in industrial tourism is because it can help to improve the image of the company and its brands. Recently, in response to an ageing customer base, Cointreau has been rebranded, targeting a much younger audience (25–50 years). Cointreau is no longer promoted as a digestive (as it used to be perceived in France) but rather as a good ingredient for cocktails. Company visits are seen as one of the creative methods to deal with the restrictions that apply to producers of alcoholic beverages. The firm invites target groups for a company visit, expecting them to become ambassadors for the 'renewed brand' (multipliers). Industrial tourism is considered particularly interesting to target the French market because the restrictions on advertisements for alcoholic drinks are more severe than in many other countries, and of course in view of distance. By the way, this does not mean that the company can ignore these restrictions during visits: the company has to be very careful with promoting its brands, but the general idea is that visitors get a more positive feeling about the product just by seeing how it is produced. Unlike in many other companies, industrial tourism is not used as a tool to attract future employees at Cointreau, notably because the image of the company as an employer is good enough. Cointreau does use company visits to show its good citizenship, particularly when it comes to issues such as 'drinking responsibly'.

In its anniversary year 1999, Cointreau received 30,000 visitors. In the last four years, however, the company has experienced declining visitor numbers: in 2007, only 22,000 people visited the plant. The falling visitor numbers also went along with a falling age average. While in 1999 the average visitor was around 55 years old, the average visitors in 2007 was only around 35 (!). Around 85 per cent of these visitors are French (estimation); the other 15 per cent mainly come from the UK, from Belgium, Germany and other countries. Most of the French visitors come from the Pays de la Loire Region, from neighbouring regions and from Paris. The decline of visitor numbers from 1999 to 2007 can clearly be related to the rebranding of Cointreau: the composition of the 2007 visitors reflects the company's target groups in a better way than the more numerous visitors in 1999 (although the share of senior visitors is still 'too high'). In addition it turns out that today's visitors spend more money, on average, in the Cointreau Shop.

Les Salines de Guérande: Terre de Sel

The salt marshes of Terre de Sel are the most visited industrial tourism attraction of the region. The visitor number increased from 42,000 in 2005 to 55,000 in 2007. The majority of the visitors (an estimated 90 per cent) is French, while most of the foreign visitors come from the UK, Germany, The Netherlands, Belgium and the French speaking parts of Canada and Switzerland (where the brand 'Sel de Guérande' is well-known). Terre de Sel is a popular excursion from the nearby beach resorts on the coast (the resort town of "La Baule" is only 3 km away), whereby Terre de Sel receives most visitors on cloudy and rainy summer days. The peak visitor period of nearby beach resorts coincides with the salt harvest in July and August, explaining why most visitors come during this period.

The visitor centre Terre de Sel was opened in 2002 by the cooperative 'Les Salines de Guérande' which represents around 190 salt workers and takes care of buying, storage, packing and marketing. The cooperative sells to supermarkets and hypermarkets, the food processing industry, restaurants and chefs, fine grocery stores, organic food shops and specialist wholesalers (Salines de Guérande, 2008). Unlike in many other places in the world, the salt production in Guérande still makes use of traditional tools and follows traditional methods that were innovated more than thousand years ago. The salt of Guérande is well known throughout France and the French speaking world, basically because many of the French first rated cooks use this salt.

The visitor centre is the starting point for guided tours around the salt marshes. Visitors can choose from 15 different guided tours, ranging from special tours for kids, families and professionals to visits with a focus on nature or on the job of the salt maker. Tours take place at fixed times and continue regardless of the number of participants (no disappointed customers), which sometimes even results in 'private tours' (tours for individuals and very small groups). One of the highlights is a 45-minute visit to the modern storage facilities; these facilities were opened to the public when Les Salines de Guèrande became member of Visit Our Companies. The storehouse-visit takes place from Monday to Friday (during working times), except in the very busy summer months of July and August. During these months, there are so many requests of visitors that Terre de Sel could only admit a very small share of interested people to this special tour; that is why they decided to suspend this special tour altogether during the summer period (again, in order not to disappoint anybody).

Terre de Sel is open every day and boasts an exhibition about salt farming and a shop where visitors can purchase the salt that is produced here and regional products that contain this salt. This shop is, however, only of minor significance for the overall salt sales of Les Salines de Guèrande: only 50 tons of salt are sold in the shop per year (at a discount price), while the overall sales of the company amount to around 7,000 tons. The sales in the shop do cover the running costs of the exhibition centre and the cost of the guided tours.

According to Terre de Sel, many of their visitors also visit STX Europe and Airbus in the nearby town of St Nazaire (a 20-minute drive from Guérande). Terre de Sel also tries to foster the links between the companies, and agreed to display information material about the tours offered by Escal'Atlantic. Terre de Sel has also made a special arrangement with local tourist offices and hotels: they display the folders of Terre de Sel and via an online booking system they can directly book a guided visit to Terre de Sel for their guests; in return they get 10 per cent of the booking fee.

INRA Angers-Nantes: Terre des Sciences

INRA is the French National Institute for Agricultural Research, the largest one in Europe (INRA, 2008). INRA Angers-Nantes is the result of a merge between the INRA centres of both cities and comprehends 23 research units located not only in the cities but in the region as well. The Angers centres are particularly specialised in horticulture (the cultivation of plants), while the Nantes centres are specialised in nutrition studies.

Our analysis of company visits is restricted to the Angers locations of INRA. In 1992, INRA Angers was involved in the foundation of Terre des Sciences which can be considered a cluster organisation for science centres in the region. One of the aims of this platform is to promote visits to companies and research centres in the field of horticulture, under the motto 'les rendez-vous de végétal'. Terre des Sciences is member of Visit Our Companies. During the spring, the summer and the autumn about 50 companies, research centres and educational institutions can be visited, resulting in an average visitor number of 1,500 a year. Over the past 10 years, more than 200 different companies have opened their doors at least once. In the programme, the timing of company visits is important: the intention is to open each company in the short period when the plants are flowering. As many companies are specialised in different plants, also the flowering period varies accordingly.

The aim of the company visits is not to increase sales (companies sell plants to wholesalers and supermarkets, not to individual customers directly), but to improve the image of the sector. The participating companies want to improve their image as employers (notably among young people) but also as good citizens. Company visits are used to create a more balanced picture of the companies, by showing the high-tech elements but also by making their activities more transparent, notably in case of genetically modified plants (a big issue in France). Moreover, for individual companies it is interesting to open their doors, because the programme generates a lot of media attention: initially only the bigger firms participated, but they were soon followed by smaller firms mainly due to the broad media coverage.

Tourist offices in the region handle the reservation of tours. Visitors have to pay a fee of €3 per tour to cover the costs made by the tourist office (not the costs made by companies). Students, pupils and handicapped persons get a discount, and there is also a card that provides entrance to five companies for €10. The duration of a visit varies between 90 and 120 minutes. The maximum group size is 30. Participating companies have to arrange a guide for the tour and they should develop a tour as well as a presentation about the company (history, finance, production, export, et cetera), following the quality guidelines formulated by Terre des Sciences.

5 Applying our frame of analysis

In this section we apply our frame of analysis for the conditions under which industrial tourism can be developed to the observations that we made in the Pays de la Loire region. This frame of analysis comprehends four groups of conditions: the selection of target groups, the attractiveness of the industrial tourism product, the benefits in terms of income and image for city and enterprise, and the organisational framework.

Target group selection

The first condition relates to the questions what visitor groups are targeted by the organisations and companies we analysed, and what groups actually visit the firms. Beginning with the first question, it appears that Visit Our Companies essentially targets companies and not visitors: their mission is to convince companies to open their doors, not to convince people to visit the companies. It could even be argued that the association believes visitors come anyway, because the demand for company visits is growing, and still higher than the supply. Although Visit

Our Companies is supply-oriented (which is also reflected in its objectives), the association is also involved in marketing the industrial tourism product, but without specifying target groups. The selection and approach of target groups is a considered a task of the individual companies and the travel agencies and tourist offices that organise tours. For instance, Escal'Atlantic, the tourist offices for St Nazaire, mainly targets schools and families (that get a discount), but also the tourists that stay celebrate their holidays on one of the beaches nearby, as well as the city trippers that stay in Nantes. For that reason the organisation has an international website and provides tours in French and English.

Visit Our Companies assists companies and agencies with target group selection by collecting visitor data in cooperation with the Regional Observatory of Tourism. A scorecard (*tableau de bord*) makes it possible to analyse the development of company visits quantitatively and qualitatively. Using this scorecard the conclusion could be drawn that the weight of international visitors is twice as large in the participating companies than in average touristic sites (for the individuals, 25 per cent of the visitors are foreigners at the members, compared to only 12 per cent on average for all sites; for groups, 15 per cent versus 7 per cent). Moreover, while all sites report a majority of groups from schools (53 per cent), one can observe that an important share of the company visits are organised by tour operators (24 per cent) and clubs for the elderly (13 per cent). The average group size is 42 persons (Charuel, Leblanc and Brin, 2006). As these data show that foreign guests (and notably the British) visit industrial sites relatively often, the association also helps its members to target these groups (e.g., by developing tours in English). The general impression is that visitors from Anglo-Saxon countries but also from China and Japan are more familiar with company visits than tourists from the European continent. Quite remarkably in view of its international orientation, Visit Our Companies' own website is only available in French, which is only partly compensated by the availability of the regional tourism committee's website (which provides information in several languages).

While Visit Our Companies targets a very broad target group (in terms of scope and scale), the annual events Made in Angers and the national open door days seem to be more oriented towards local citizens, and less towards tourists (although, of course, companies are open to everyone). More in particular it seems that these events aim to attract young people in order to promote the companies as employers. In practice, however, it appears that events are also very popular among retired and unemployed people, who obviously have more time to visit firms than people with a job. Made in Angers reports that about 50 per cent of its visitors are retired people, while only 17.5 per cent falls in the target group of students and pupils. In general it appears to be difficult to attract the group between 25 and 40 years old, which is also confirmed by the visitor statistics of the Autonomous Port of Nantes-St Nazaire.

In general individual companies seem to define the target groups of company visits more narrow than tourist organisations. They are particularly interested in primary stakeholders (customers, suppliers, business partners, etc.) and less in the general public. This can be illustrated by the strategy of Cointreau to invite young people who fit in the new branding strategy for the liqueur. According to Cointreau it is better to receive 10 people who match with the desired profile of customers than 20 people who don't. Young people are also clearly targeted by the companies under the umbrella of Terre des Sciences. The Autonomous Port of Nantes-St Nazaires seems to target inhabitants of the region, and again young people in particular (as potential employees), but in practice Escal'Atlantic takes a broader approach also targeting (foreign) tourists. This might not be in line with the objectives of the port, but for the tourism agency it is simply a matter of gaining income. A conclusion could be that the outsourcing of company visits can result in a mismatch between the desired target groups and the groups that actually come in. For Airbus and STX Europe the leisure-oriented visitors that come in through Escal'Atlantic are less interesting – from a business and marketing point a view – than the business and education-oriented visitors they take care of themselves (in which schools and customers can participate for free). Terre de Sel is one of the exceptional examples of an individual industrial tourism supplier that targets a very broad audience, simply because each individual is a potential customer. The needs of different groups have been anticipated by developing 15 (!) different tours.

Attractiveness of the industrial tourism product

The second condition for the development of industrial tourism concerns the attractiveness of the industrial tourism product, which partly depends on its accessibility. As stated above the activities of Visit Our Companies are strongly oriented towards the improvement of these conditions. Instruments used to that end are quality control (participation in a national quality programme), workshops and a website with links to websites, contact persons and telephone numbers of the members. While quality assessments are mainly used to secure the basic requirements for receiving visitors (parking facilities, reception, toilets, signing, safety, transparency etc.), workshops, often in combination with visits to proactive companies, provide opportunities to exchange knowledge on how to make the products more attractive. The association refers to the experiences of regional forerunners in the field of industrial tourism to make other companies aware of the possibilities to improve their offer. Of course, some firms are in the fortunate position of having a very attractive product and production process (e.g., Airbus, STX Europe, Cointreau), but that does not imply that ideas cannot be transferred from one company to another.

Looking at the experiences of the companies that we analysed in detail, several lessons can be drawn concerning the question how to make company visits more attractive. In our sample Cointreau is the company that invested most in the attractiveness of its industrial tourism product, anticipating the growing 'experience economy'. The elevated walkway not only allows visitors to see the bottling process, but is also an integrated part of the Cointreau museum which provides a full overview of the company's history with old advertisements for the product and even an overview of Cointreau imitations from all over the world. This combination of a museum and a visit to the operational plant makes the tour very attractive. Another attractive element of the Cointreau tour is the confrontation with the product itself, notably because the visitor is invited to mix their own cocktail and the appealing environment in which this takes place (a trendy bar). The relative short employment period of the professional and well trained guides (most guides only stay for one year) nicely fits in the company's strategy to keep the tours attractive: if they do the job for too much time they might become too 'automated' and lose some of their enthusiasm. Moreover the company prefers young guides in order to convey a young and dynamic image. It is, however, not a policy to send guides away after one year; in most cases the guides take this decision themselves.

The attractiveness of the visits to Airbus, STX Europe and the Autonomous Port of Nantes-St Nazaire is mainly the responsibility of the tourism organisation Escal'Atlantic that organises these tours. Escal'Atlantic takes care of the basic requirements to make visits attractive such as headphones, information displays on the site and professional guides who have up-to-date knowledge about the company. Visitor toilets are not available on all sites for the simple reason that such visitor facilities are available at the office of Escal'Atlantic from where all tours depart. The three tours are easily accessible thanks to the visibility on the international website of this tourism office. St Nazaire presents the company tours as one of the highlights of the city, more or less on the same level as the two more traditional tourism attractions (the ocean liner museum and the submarine espadon), that also fit in the economic profile of the city. Escal'Atlantic is in the favourable position of having access to three attractive firms that obviously appeal to the general public; this particularly applies to STX Europe since it is one of the few places where cruise ships are produced. The St Nazaire sites of Airbus are considered slightly less attractive than STX Europe, notably because the final assembly does not take place here but in Toulouse (and Hamburg). The port is the least visited industrial tourism attraction offered by Escal'Atlantic: it is not as unique as the other two sites but it is well accessible: visits can be made either in a group by coach or by private modes of transport (car, bike, etc.) using a mobile phone application.

Terre de Sel is the most popular industrial tourism attraction of the region. Visits to the salt marshes can be seen as a combination between visiting operational firms and heritage tourism since the marshes have been there for centuries and the production method hasn't changed at all. Moreover the location, near the beaches of the Atlantic coast, is very favourable for attracting tourists. Terre de Sel, however, also owes its attractiveness to the policy that customers should not be disappointed: tours are never cancelled and during the summer months the popular storehouse can simply not be visited to avoid a situation in which the supply cannot meet the demand. Quality is clearly a leading principle.

Benefits in terms of income and image for city and enterprise

The third set of conditions concerns the benefits in terms of income and image for city and enterprise. According to Visit Our Companies there is a clear difference between small and large companies. For most large companies (e.g., Airbus, STX Europe, the Autonomous Port of Nantes-St Nazaire, Cointreau) industrial tourism is mainly a marketing and communication instrument, while smaller companies see it is an opportunity to earn some extra money. Particularly companies in the food and beverages sector use industrial tourism to sell some of their products; in the case of Terre de Sel this extra income is, however, very modest with not even 1 per cent of the total sales being generated by the shop. Under the umbrella of Terre des Sciences small companies in the horticulture sector seem to consider industrial tourism as a marketing instrument rather than an additional source of income. The three big companies in St Nazaire receive some income through the reception of leisure visitors via Escal'Atlantic; for example, STX Europe earns €4,000 a year, because the tourism office compensates the company for the right to organise tours. However, in comparison with the annual turnover of these companies, the direct financial benefits are negligible.

In general we observe that industrial tourism is being used as a means to bridge the gap between society and business. Not only big firms but also smaller firms that feel pressure on the licence to operate (e.g., companies under Terre des Sciences) open their doors to present themselves as good citizens, trying to correct the images and perceptions that live in the minds of people. For large companies such as the Autonomous Port of Nantes St Nazaire company tours are particularly important to communicate their added value for the region's economy. They expect that participants in company tours can become ambassadors for the port, sharing their experiences with numerous friends and relatives. In addition, nearly all companies we analysed expect labour market benefits from industrial tourism. They see company tours as an instrument to promote themselves as attractive employers, trying to take away prejudices about the sector they work in. The only exception is Cointreau since this company experiences no problems with finding adequate labour; for Cointreau industrial tourism is more an instrument to approach customers, replacing traditional advertisements that suffer from stricter rules on a national level. Airbus and STX Europe also welcome end users (passengers of cruise ships and aeroplanes), but mainly because it is in the interest of their customers: visits to these companies promote the use of these modes of transportation, for instance by reducing the fear of flying in the case of Airbus. Not surprisingly, both companies expect more benefits from visits by direct stakeholders than from visits by indirect stakeholders.

Not only companies expect to benefit from industrial tourism, also regional and local authorities recognise its added value. Company visits may not bring in millions of visitors, but they fit in a more comprehensive strategy to promote 'economic discoveries' which also includes visits to industrial heritage and science tourism. Several studies confirm that a small but substantial share of the visitors combines visits to traditional attractions with visits to companies. Particularly companies located in the coast area (in the towns of St Nazaire and Guérande) seem to attract beach tourists, notably on cloudy, rainy days. For St Nazaire – and its tourism organisation Escal'Atlantic – industrial tourism is an interesting niche market since the town has no attractive historical city centre. The organisation of company visits is profitable, although the margins are relatively low. The actual benefits, however, are considered higher because the company visits also contribute to an improvement of the city's image. For the City of Angers, Cointreau has been one of the major tourist attractions for a very long time. In both cities, companies are not the primary tourist attractions, but they do contribute to the diversity of the tourism supply.

Industrial tourism not only benefits the participating companies but also the region in which they are located. Visitor statistics demonstrate that industrial tourism is a small but substantial segment of the regional tourist industry. Company visits are not only popular among 'urban tourists' (city trippers) but also among families who spend their holidays on a camping site near the Atlantic coast. Moreover, company visits also help to promote the region as an alternative for Paris, as pride citizens and visitors become ambassadors for the region and thus possibly strengthen the West Coast Effect. In the City of St Nazaire, industrial tourism is considered a strategic element of a more comprehensive strategy to improve the image of the city, in combination with a waterfront redevelopment project. For Angers, the main motive to develop industrial tourism has been to strengthen the relation between business and society. The initiators of Made in Angers expect that this event can result in better linkages between educational institutions and companies (preparing people better for specific jobs) which in the end helps to improve the performance of the regional economy.

The experiences in the Pays de la Loire region not only demonstrate that city and enterprise expect to benefit from company visits, but also that their interests are in some aspects converging. All actors involved have an interest in positive media coverage about the initiatives (facilitated by Visit Our Companies and Made in Angers, among others), and a better image of the industries in the region. Also the expectation that company visits can help to improve the relation between business and society is an example of converging interests.

Organisational set-up

The fourth condition is the organisational set-up which refers to the internal organisation within companies as well as the organisational framework on a local or regional level.

The three companies in St Nazaire (Airbus, STX Europe and the Port) have in common that they have all outsourced the reception of leisure visitors to Escal'Atlantic. This tourism office is not only responsible for reservations, guides and transport, but also takes an active part in making the sites more attractive, by investing in walkways, information displays, signs, etc. This demonstrates that

Escal'Atlantic has a strong financial basis, with financial support from public and private shareholders. Visitors with a business motive or students from universities and technical schools are in most cases received by a department within the company: this can be a communications department (Port, STX Europe) or a human resources department (STX Europe) or a specialised department for visitors (Airbus). In Angers we saw that Cointreau also has its own visitor services department, which belongs to the marketing and communication division. These examples show that the internal organisational set-up of industrial tourism is related to the main motives: general communication (marketing) and attracting human resources.

On the level of the region, the association Visit Our Companies has developed an environment in which other actors – companies, tourism offices, tour operators – can create an attractive and accessible industrial tourism product. With only one permanent staff member, management costs are kept low. Interestingly, Visit Our Companies does not really 'push' companies to become member: this is considered to be a loss of energy. The idea is that companies have to be intrinsically motivated to become member. The association's role is not only to recruit companies, but rather to select companies that meet the high quality standards. Furthermore, the association is clearly a platform for the private sector, which is particularly reflected in the composition of the board of managers, and the fact that big companies such as Airbus and Electricite de France (energy company) have delivered the president of the association so far. The commitment of member companies is relatively high, with about 60 per cent of the members showing up at the annual meeting of the organisation. The Chamber of Commerce and Industry has assumed the role of facilitator, while the regional committee of tourism is taking care of integrating 'economic discoveries' (including company visits) in the regional tourism strategy. Apart from quality control, the association also promotes the exchange of knowledge and experiences; the activities of active big firms such as Airbus, STX Europe and Cointreau are used to inspire other firms, notably small and medium sized companies. This also explains why membership fees for big companies are relatively low: large companies are expected to take a more active role in the association, for instance by giving presentations to foreign delegations. The active membership of large firms gives the association a more powerful position, towards other companies as well as politicians.

In Angers we have observed that industrial tourism not only implies the involvement of companies (and the Chamber of Commerce of Industry), but also of educational institutions. The University of Angers (ESTHUA) has played a key role in the development of 'economic discoveries', with its involvement in the first international conference on company visits, the Made in Angers event and the creation of a bachelor's programme dedicated to this sector. In all initiatives we observe strong cooperation between the various sectors: for instance, lectures in the bachelor's programme are given by academics as well as representatives from the corporate world. Not coincidentally, the lecturer who is the driving force behind the establishment of the course was also involved in the foundation of the association Visit Our Companies, confirming the importance of leadership.

References

Airbus 2008. [Online]. Available at: http://www.airbus.com [accessed: 2008].

Autonomous Port of Nantes-St Nazaire 2007. *Annual Report.*

Autonomous Port of Nantes-St Nazaire 2008. [Online]. Available at: http:// www.nantes.port.fr [accessed: 2008].

Charuel, C., Leblanc, J. and Brin, O. 2006. Visitez les entreprises en Pays de la Loire!, In J.R. Morice, ed. *Visite d'entreprises; Actes du premier Colloque Européen de la Visite d'entreprise*, Les Cahiers Espaces, 92, pp. 50–53.

Cointreau 2008. [Online]. Available at: http://www.cointreau.nl [accessed: 2008].

INRA 2008. http://www.international.inra.fr/ [accessed: 2008].

MarketNewZealand.com 2008. *Expanding Cruise Ship Market.* [Online]. Available at: http://www.marketnewzealand.com/MNZ/news/story/14401/19446.aspx [accessed: 18 July 2008].

Nantes Island 2008. [Online]. Available at: http://www.iledenantes.com. [accessed: 2008].

Observatoire Régional du Tourisme en Pays de la Loire 2007. *Chiffres-clés du Tourisme en Pays de la Loire.*

Salines des Guérande 2008. [Online]. Available at: http://www.salinesdeguerande.com/ [accessed: 2008].

SEM Régionale des Pays de la Loire 2007. *Profil des touristes des sites majeurs en Pays de la Loire.*

Terre de Sel 2008. [Online]. Available at: http://www.terredesel.fr [accessed: 2008].

Ville Port 2008. [Online]. Available at: http://www.mairie-saintnazaire.fr/pages/autres/ville-port-2.php [accessed: 2008].

Visit Our Companies 2008. [Online]. Available at: http://www.visiteznosentreprises.com [accessed: 2008].

Discussion partners

Mr. Emmanuel Blanc, Managing Director of Terre de Sel.

Mr. Philippe Broix, Director of Angers Tourism office.

Mr. Christian Charuel, President of the Association 'Visitez nos entreprises en Pays de la Loire'.

Mr. Sylvain Dourneau, President of the Tourism Department of the Chamber of Commerce and Industry of "De Maine et Loire".

Mr. Daniel Dupuis, President of the Tourism Committee of Pays de la Loire.

Mr. Pascal Frenau, Communication Manager of the port authority of Nantes-St Nazaire.
Mr. Jean-Luc Gaignard, Communication Manager, Terre de Sciences/INRA.
Ms. Catherine Houdet, Director of the Tourism Department of the Chamber of Commerce and Industry of Pays de la Loire.
Ms. Isabelle Huyghe, External Communication Manager, STX Europe.
Ms. Julie Leblanc, 'Visitez nos entreprises en Pays de la Loire'.
Mr. Emmanuel Mary, Director of Escal'Atlantic.
Mr. Jean-Rene Morice, Lecturer at University of Angers.
Ms. Valerie Quinaud, Communication Assistant, Remy Cointreau.
Mr. Philippe Rettig, Communication Manager, Airbus.

Chapter 5
Industrial Tourism in Turin

1 Introduction

In this chapter we analyse the development of industrial tourism in Turin and the Piedmont region. The case of Turin was selected because the tourism organisation – Turismo Torino e Provincia – and the Chamber of Commerce have jointly developed a special industrial tourism programme (Made in Torino; Tour the Excellent) which targets visitors and residents. Moreover, Turin is a typical industrial city that is trying to convert itself into a modern European city with a more diversified economic structure. It is interesting to see how Turin is coping with the challenge of improving its image without neglecting its industrial base. Not surprisingly, the Olympic Games in 2006 have played a key role in the development of tourism in general, and industrial tourism in particular.

This chapter is structured as follows. Section 2 presents a general profile of Turin, with a clear focus on the tourism supply and demand. After that, Section 3 deals with the organisational structure of industrial tourism, introducing the main stakeholders on city and regional level as well as their policies and actions. In the fourth section we discuss the activities of three companies – Fiat, Gufram and Aurora – to see what actions have been taken on the level of individual firms. The fifth section confronts the findings with our frame of analysis that we developed in Chapter 2, resulting in some conclusions with regard to the conditions under which industrial tourism has been developed.

2 General profile of the city and the region

With 900,000 residents (2007 data), Turin is the fourth largest city of Italy (after Rome, Milan and Napoli). The city is located at a distance of 80 km from the French border and 100 km from the Swiss border. It has a modern regional airport with more than 30 destinations in Europe (SAGAT, 2006) and more than 3 million passengers a year. A high-speed train connection between Turin and Milan's international Malpensa airport (which was completed in 2009) will further improve Turin's international accessibility. Travel time between Turin and Milan has been reduced from 90 to 50 minutes. Turin's accessibility could be enhanced by a high-speed train connection with Lyon, but this project is topic of debate, notably because of the environmental impact on the Alps.

Turin is by far the largest municipality of the province of Turin, which is the most fragmented province of Italy with no less than 315 *comuni* (municipalities).

The province of Turin is one of the eight provinces in the Piedmont region, one of the 20 autonomous regions in which Italy is divided. This region, which borders with France and Switzerland, covers an area of more than 25,000 km^2 and counts approximately 4.3 million inhabitants.

Economic and demographic context

In 29 B.C. Turin was founded as a colony of the Roman Empire. In the first millennium the town was part of several other empires before it became an autonomous city-state in the 12th century. In 1280, the city was captured by the dukes of Savoy, marking the beginning of a new era in which Turin developed into an important centre of artisan manufacturing, with a specialisation in textiles. In the 18th century Turin became an important centre of trade and finance and it was chosen as the first administrative capital of the unified Italy in 1861, a status it lost however to Florence only four years later (Rome became the capital in 1871). From then on the city started to invest in developing new industries, with the car manufacturing industry as one of the pillars. In 1911 a third of the city's expanding workforce was employed by the automotive sector. In the decades that followed the city's flourishing economy was accompanied by a massive inflow of immigrants, notably from the south of Italy. Between 1921 and 1939 the population grew from almost 500,000 to almost 700,000. After the Second World War the city recovered fast, mostly because 38 per cent of the Marshall Plan Funds earmarked for the Italian engineering sector had been invested in Fiat. With the expansion of the automobile market Turin became even more dependent on its car manufacturing industry: at the end of the 1950s approximately 80 per cent of all industrial activity was related to the production of cars. Between 1951 and 1971 Fiat expanded its workforce in Turin from 47,000 to 145,000 (including 30,000 white-collar workers). The growth of the population further continued, reaching a peak of 1.2 million in 1975.

The period of economic growth came to a halt in 1973 with the global oil crisis. It soon turned out that Turin had become too dependent on one industry and one company in particular. For several decades the share of industry in local employment had remained fairly high, because declines in the textile, clothing and food production industry were offset by increases in the automobile industry. Fiat, started to shift production away from Turin, mainly to the struggling south of Italy, with financial incentives of the State. The relation between the city and the (automobile) industry also suffered from the tensions between workers and the company resulting in strikes, anti-capitalist protests and even terror movements.

In 1980, the metropolitan area lost 100,000 industrial jobs (Maggi and Piperno, 1999). In several stages Fiat reduced its workforce, with 23,000 jobs in 1980, and 45,000 jobs between 1986 and 1996 (Rosso, 2004). Without the support of the Italian State, in the years before the European Union prohibited such trade protections, Fiat would probably not have been able to survive in this period. The share of Turin plants in the production of Fiat dropped from 60 per cent in the early

1990s to less than 30 per cent in 2002. Currently, Fiat only employs about 30,000 people in the city (Symcox and Cardoza, 2006).

With the rise of the service sector, the share of industry in Turin's economy has decreased from more than 60 per cent in 1971 to less than 40 per cent in 2002. In 2005, the share of industry in the active population was 34.4 per cent, while private services took a share of 44.4 per cent. The share of industry is, however, still higher than in many other Italian and European cities. The city now has a more diversified economic structure. The automotive industry is no longer the only key sector in Piedmont: other important sectors are aerospace, ICT, life sciences and biotechnology, logistics, nanotechnology, environment and renewable energy, artistic handicraft, home textile, contemporary art, writing tools, agro-food and the movie industry (Invest in Torino Piemonte, 2008; Chamber of Commerce for the Turin Province, 2008). In terms of export, however, the automotive industry is still the most important sector, accounting for 40.5 per cent.

Tourism profile

The Piedmont region has a long history of tourism – the old spa town of Acqui already attracted visitors two thousand years ago – but it is not a typical tourist destination. In comparison with other historical cities (like Rome, Florence and Venice) and coastal towns (like Rimini) Piedmont is a latecomer to tourism (Piedmont Region, n.d.). Geographically, the tourism product of the Piedmont region can be divided into four sub-areas: the Olympic mountains (skiing and hiking), the lake area (including Lago Maggiore), the hills (including Langhe), with wine, castles (some of which are royal residences) and culture, and the City of Turin, an important MICE (meetings, incentives, conferences, exhibitions) destination.

For a long time the city of Turin and the region of Piedmont paid more attention to industrial development, but like many other industrial cities, Turin has discovered tourism as a growth cluster that has the potential to compensate for the job losses in the manufacturing industry. Many investments in tourism in the last decade can be related to the strategic plan (*Piano Strategico*) 2000–2010 which formulates 20 objectives and 84 actions, categorised in six strategic lines (City of Turin, 2000). The strategy is based on a vision that presents Turin as a European metropolis that is ready to compete in the knowledge economy. The plan has been developed in consultation with universities, industry, banking, foundations, and other relevant stakeholders such as the general public (with public meetings).

One of the six strategic lines is to promote Turin as a city of culture, tourism, commerce and sport. Another is to promote enterprise and employment. With the loss of jobs in industry, more and more attention is being paid to the development of other sectors, tourism being one of them. The Winter Olympics of 2006 have been used by the city to realise many of these objectives, but notably to improve accessibility (another strategic line), to diversify the economy and to promote the city as a tourist destination and business location (Van den Berg et al. 2002). One of the objectives is to position the destination of Turin and the Piedmont region in the international tourism market. Before the Olympic Games took place, Turin clearly had an image problem in the international market. Most people had no image of the city whatsoever, while the remaining minority mainly associated the city with Fiat and the (car) manufacturing industry, the image of the city being 'still very much tied to its industrial past' (City of Turin, 2000, p. 99). Two actions have been defined to improve the image: 1) to create and expand the Turin and metropolitan area product into the national and international tourist market; and 2) to create a 'committee' to spread the wine and food culture of both the city and region into a national and international arena.

With the Olympic Games as excuse – being inspired by the Barcelona 1992 strategy – Turin and Piedmont have invested heavily in tourism development. A special regional law (law 18) was created in 1999 to provide public funds for private investments in the tourism industry. Investments in tourism increased from about €174 million in 2000 to almost €409 million in 2002, resulting in a considerable expansion of the hotel capacity. In 2005, Turin had a supply of 10,310 beds in hotels and residences, of which 4.8 per cent in five-star hotels and 35.3 per cent in four-star hotels, and another 3,930 beds in other accommodations (like: bed & breakfast, camping, youth hostel, and so on) (City of Turin, 2005). Another regional law (law 4), passed in 2000, enabled the public and non-profit sector to invest in tourist areas, leading to investments worth €196 million.

On the demand side several sources confirm that Turin and the Piedmont region have been successful in attracting tourists. According to European Cities Tourism (2007), bednights in the City of Turin increased with approximately 50 per cent between 2000 and 2005, indicating that the Olympics already paid off before the event had taken place. More recent statistics demonstrate the growth in bednights between 2000 and 2005, with a continuation in 2006, the year in which the Olympic Games were held, to the record number of 3.3 million bednights. It also becomes clear that foreigners took a large share in this increase. In 2007, the year after the event, the number of bed nights in the metropolitan region of Turin decreased to a level only 10 per cent higher than the 2000 figure (2.43 million compared to 2.2 million), which is however quite common when cities host a mega event. In 2007, the total number of bed nights registered in the region was 10.3 million. The Province of Turin accounted for 38 per cent (more than 3.9 million bed nights), while the metropolitan region (a smaller area than the Province) took a share of about 24 per cent. The share of foreigners in the number of bed nights registered in the Province is relatively modest with less than 26 per cent, with the UK (5.3 per cent), Russia (4.0 per cent), Spain (1.9 per cent) and France (1.8 per cent) as main countries of origin. Other parts of the region are more successful in attracting tourists: the Lake District (surrounding) counted 3.3 million bed nights, with a share of foreigners of more than 71 per cent. As is common for cities, most tourists arrive during springtime and in the autumn (Sviluppo Piemonte Turismo, 2007; City of Turin, 2005).

Turin has a very broad and large tourism supply. The historical city centre is an attraction in its own right, with various squares, churches (for example the Church of San Domenico and the Cattedrale di S. Giovanni Battista) and palaces as highlights. In 1997, the royal residences in Turin and its region – including Palazzo Reale and Palazzo Madame – were declared UNESCO world heritage sites. In the near surroundings tourists can visit an impressive amount of castles (of Rivoli, Moncalieri and Agliè) and royal residences such as the Reggia di Venaria Reale, a summer residence of the Savoy family. Visitors have access to numerous museums, with the National Museum of Cinema and the Egyptian Museum (the second largest after the one in Cairo) as top attractions, with more than 300,000 visitors in 2004. Also worth mentioning is the Car Museum (*Museo dell'Automobile*) which attracted more than 65,000 visitors in 2004 (City of Turin, 2005).

After the Olympic Games in 2006 Turin continued its strategy to organise many events (again following the Barcelona example). The city succeeded to become the first World Design Capital being an occasion for numerous exhibitions, conferences and festivals in 2008. One year later, in 2009, the city hosted the European Indoor Championships in athletics. On a more regular basis the city hosts several fairs, some of which are clearly related to the key sectors in Turin like the Torino Film Festival and Salone del Gusto (Taste Festival). Remarkably, Turin is no longer host of an Auto Show: the *Salone dell'automobile di Torino* was held for the last time in 2000.

3 The organisational framework for industrial tourism in Turin

In this section we present the organisational framework for (industrial) tourism in Turin and the Piedmont region, introducing the main stakeholders and their strategies.

Tourism organisations

In 1993 it was decided to make the 20 autonomous regions in Italy – Piedmont being one of them –responsible for their own tourism development and promotion. One of the regional boards (*direzione regionale*) deals with tourism, trade and sports. One of the sectors (*settori*) of this board has the task to promote the Piedmont region as tourism destination, covering all tourism products: the mountains (for skiing and hiking), the lakes (like Lago Maggiore), wine producing areas, food production, cultural heritage, industrial heritage (Lingotto), royal residences (La Venaria Reale) and gastronomy.

Turismo Torino e Provincia Within the region nine local tourism agencies are nowadays operational (there used to be 12). One of them is Turismo Torino e Provincia (Turin Tourist Board). This agency has recently been created by merging three local tourism organisations, including Turismo Torino which used to be responsible for the tourism promotion of Turin only. In the new setup Turismo Torino e Provincia[1] is the largest local tourism organisation of Piedmont, and even of Italy. 'The Consortium originated on the one hand from observing that Torino – as capital of Piedmont and as a city now known worldwide thanks to the Olympic Games – can play a leading role for the whole province; on the other hand from the will to recover and strengthen the relationship between Torino and its mountains, between the city and its crown-like Savoy Residences, between the regional capital and the little towns that surround it, between the Italian capital of industry and innovation and the food-and-wine traditions of the province.'[2]

The new tourism organisation for the Province of Turin is partly financed by public bodies (the City of Turin, the Province, the Region, the Chamber of Commerce and the other municipalities; together they own 95 per cent of the organisation) and private actors (the tourist industry). The organisation operates three information offices in Turin – one in the city centre, one at the central station, and one at the airport – and fourteen in the Province. The organisation has developed several products such as the Torino+Piemonte Card (providing free access to cultural attractions and public transport as well as several discounts on other services), Gourmet in and around Torino (special menus at restaurants), and Chocopass (a set of coupons for tasting chocolate). In addition the tourism agency organises several guided tours, such as a city tour and a visit to the Egyptian museum.

Sviluppo Piemonte Turismo

In 2007, a special research and development agency for tourism in the Piedmont region – Sviluppo Piemonte Turismo – was set up in order to increase the impact of tourism on GDP. This agency is a public company, 100 per cent owned by the region. Its foundation can be related to the Olympic Games that helped to put tourism on the political agenda. The organisation has five areas of activities: 1) market research and statistics (qualitative and quantitative analysis), particularly counting overnights because this data is used by the central government for the distribution of funds, but also studies on specific issues such as customer satisfaction and impact studies; 2) strategic marketing based on research and statistics, focusing on strong products with a high added value for the region such as food and wine, biking, and culture, for instance by developing packages with airlines; 3) product development (differentiation) in cooperation with local tourist boards and incoming tour operators, and training because people make the difference; 4) presenting all products of the region on the joint website: www.torinopiemonte.com and 5) administration and logistics (e.g., of promotional materials).

1 Hereafter abbreviated as Turismo Torino.

2 www.turismotorino.org.

Made in Torino; Tour the excellent

In 2005 Turismo Torino, before it merged with the two other local tourism organisations, took the initiative, together with the Chamber of Commerce, to develop an industrial tourism programme named 'Made in Torino; Tour the excellent'. The idea was to use the Olympic Winter Games as a catalyst to promote the attractiveness of Turin, not only as a tourist destination, but also as a strong economic region. Industrial tourism was considered an instrument to improve the image of Turin, as before the Olympics the image of the city abroad used to be weak. Many people only knew that Turin is the home base of Fiat, but obviously, Turin has more to offer. From a tourism point of view the region has a very attractive product, with the mountains nearby, things to see in the city, good food, and so on. Also from an economic point of view, the city is more than Fiat, with several other global companies such as pen producer Aurora and design companies like Giugiaro and Pininfarina. Industrial tourism is used by the initiators to show the excellence of Turin and its companies to the world: it is an addition to the rich tourism product of the city. The programme not only targets visitors – mainly second or third time visitors who want to see something different – but also residents who can act as ambassadors of the region, making the rather modest population of the region a bit more proud of their assets, an objective which was also aimed for by hosting the Olympic Winter Games. Hence, 'Made in Torino' is more than a tourism campaign: it is a joint effort of tourism and business to improve the image of the city as tourist destination and business location, notably in view of the Chamber of Commerce's participation.

Since 2005 Turismo Torino and the Chamber of Commerce have step by step developed a package of tours to companies in the region. They started in 2005 with the automotive sector (Fiat being the first company to be included), also including car three designers, and the pen producing sector. In 2006, the programme was further enriched by adding visits to furniture designer Gufram (as part of the 'design' sector) and to several companies that operate in the aerospace sector. In 2007 also New Holland and Iveco (automotive) joined the programme. In 2008 producers of coffee, beverages, bread, chocolates and sweets were invited as representatives of the 'taste makers' sector, with famous names such as Martini and Rossi and Lavazza. In the years to come there are possibilities to further expand the programme, for instance by visits to research centres (for example biotechnology) or multimedia companies, or other companies that belong to sectors listed by the Chamber of Commerce as key sectors. Another innovative idea is to show some products of the participating companies at Turismo Torino's information office in the centre, but that would require a relocation of the office.

For each tour about 25 people can participate. Initially the tours could only be booked over the internet, which made it relatively difficult for tourists to buy tickets: in many cases, tours were already sold out weeks before the tour took place. In practice, many tourists do not book factory visits weeks or months in advance; they prefer to do this during their stay. To solve this problem, it was decided in 2008 to share the availability of tours between the internet (10 to 15 places) and the Turismo Torino call centre. In the first year of its existence all tours of the programme were fully booked, with about 400 visitors participating in 24 tours. The majority (an estimated 70 per cent) came from the region, 20 per cent from Northern Italy, and 10 per cent from abroad (including Germany, The Netherlands and Finland). About 60 per cent of the visitors are more than 50 years old, 23 per cent between 36 and 50, 9 per cent between 26 and 35 and 5 per cent between 16 and 25. The minimum age for visitors is 16, which is because of security reasons (and the associated insurance costs). Male visitors account for 59 per cent. Interestingly, citizens tend to do the same tour for several times, maybe also inviting friends and relatives. Statistics also demonstrate that women tend to favour other tours (Gufram, Galup, Aurora, Thales Alenia Space) than men do.

In 2008 there were 40 tours to 20 companies in total (so, with an average of two visits per company). All tours take place on Friday, in the morning or the afternoon (half a day). This combines the aim to attract people who stay in Turin for a (long) weekend and the wish of visitors to see factories in operation (thus making Saturday and Sunday less appropriate). All buses depart from the city centre – in front of Turismo Torino's information centre – and usually visitors get to see a video in the bus to the factory, or there is a general presentation. Because of accessibility restrictions participating companies have to be located not too far from the city of Turin. In most cases visitors get to see the company itself, except for the three car designers; on the website of Turismo Torino it is stated that 'during the visits of Bertone, Giugiaro and Pininfarina you will not see production sites' for the very simple reason that car designers are not fond of giving competitors a view on the design process. Instead visitors enter a museum.

Participating companies do not pay, but also do not receive any money. Visitors pay €8 to €10 for each visit, which is spent on transportation, insurance and staff (hostesses provided by Turismo Torino). The income from ticket sales, however, is by far not enough to cover all costs. Or to put it otherwise: the prices of the

Table 5.1 Sectors and companies in the Made in Torino tour

Sector	Since	Companies
Automotive	2005	Fiat, IED, Iveco (2007), New Holland Kobelco (2007)
Design	2005	Bertone, Giugiaro, Pininfarina, Gufram (2006)
Pens	2005	Aurora, Lecce Pen, Stilolinea[1]
Aerospace	2006	Alenia Aeronautica, Thales Alenia Space, Galileo Avionica
Taste makers	2008	Caffarel, Galup, Caffè Vergnano, Lavazza, Martini and Rossi, Pastiglie Leone, Associazione Panificatori della Provincia di Torino

Note

1 Stilonelinea decided not to participate in 2008.

tickets are too low. To close this gap in the budget, the Chamber of Commerce co-finances the initiative with an annual subsidy that mounted from €60,000 in 2006 and 2007 to €100,000 in 2008. The main motive is a strong belief that industrial tourism will help to further strengthen the economic structure of the region, being an instrument to show excellence to the world. The Chamber sees its participation in industrial tourism as strategic, long-term investments that fit in its multi-year strategy. Staff costs of the Chamber of Commerce are not included in the amounts mentioned above, implying that the contribution of the Chamber is actually even higher (an estimated 20 to 25 per cent). The Chamber takes an active part in the organisation of the programme, particularly through involvement in the selection of themes and identifying companies that could open their doors. Three clusters (automotive, design and aerospace) that have been included are 'centres of excellence and manufacturing specialisations' (Chamber of Commerce, 2005), while also the production of (agro-)food and writing tools have been recognised as key sectors. The Chamber of Commerce – a public organisation with legal tasks – thus acts a link between Turismo Torino and the companies, since many companies also participate in its other initiatives.

Although the programme is fully organised and funded by the initiators it has not been easy to involve the companies, and without the Chamber of Commerce's support it would probably have been impossible. Many firms are too busy with their core business activities, being concerned about increasing global competition and paying little attention to opening their doors for visitors. Some firms, such as Aurora and Fiat, have broadened their view and realise that industrial tourism should be at the core of their business: to communicate their brands and to secure their license to operate. Visibility and transparency – showing excellence in many aspects – are important motives for these firms to participate. Many companies that participate in the 'Made in Torino' tour also open their doors for visitors through other channels, targeting their stakeholders (customers, employees, suppliers, financers, and so on). The only sector that has always had a strong interest in attracting tourists – not only in Turin, but probably everywhere – is the food and beverages sector (like wine production and chocolate tasting) as tourists are expected to spend their money on buying their products on the spot.

Made in Torino targets companies in the Turin Province with a focus on the Turin metropolitan area, because of accessibility. On a regional level, industrial tourism is mainly associated with food and wine production and industrial heritage, and not so much with visits to operational firms. Regional tourism organisations consider industrial tourism as a niche market that lacks the potential to bring in large amount of visitors, and since their budgets from the national government are related to the number of bed nights (a performance indicator) they have few incentives to invest in such niche markets. Moreover, it seems that decision makers on a regional level are somewhat reluctant to use the industrial base as an asset for tourism, rather preferring 'not to talk about it'.

4 Turin's industrial tourism products

In this section we discuss the industrial tourism activities of three companies in more detail: Fiat, Gufram and Aurora. This small selection covers three of the five sectors included in the programme Made in Torino: automotive, design and pens.

Fiat

The importance of car manufacturer Fiat (Fabbrica Italiana Automobili Torino) for Turin and the region needs little explanation: the history of the company and the city are strongly interrelated, as discussed earlier. The city has always been very vulnerable for the ups and downs of the company, but today the outlook is rather positive with the new 500 model as an image builder, not only for the company but for the city was well.

Fiat was founded in 1899. Since then the Fiat Group developed into a global company producing automobiles of several brands (Fiat, Fiat Professional, Alfa Romeo, Lancia, Maserati and Ferrari) as well as trucks and commercial vehicles under the brand of Iveco and agricultural and construction equipment under the brands of Case New Holland (Case, New Holland, Kobelco). Other activities are the development of components and production systems and publishing (like the daily newspaper La Stampa). In total the Group employs more than 185,000 people worldwide (Fiat Annual Report, 2007). More than 40 per cent of the workforce is located in Italy, a considerably higher share than the percentage of revenues generated by the Italian market (27.1 per cent). We already mentioned that the number of Fiat employees in the Turin province is approximately 30,000 (Galasco, 2007 in Winkler, 2007).

Fiat plays a key role in Piedmont's automotive industry. According to Invest in Torino Piemonte (2008) – the inward investment agency for the region – this cluster comprehends over 1,400 companies employing more than 75,000 employees. Among these companies are renowned car designers such as the Bertone Group, Italdesign-Giugaro and Pininfarina. The automotive clusters also

Table 5.2 Selected data by region

	Employees		Revenues	
Italy	77,679	41.9%	15,857	27.1%
Europe excluding Italy	45,999	24.8%	23,461	40.1%
North America	11,364	6.1%	5,842	10.0%
Mercosur (South America)	39,324	21.2%	8,318	14.2%
Other regions	10,861	5.9%	5,051	8.6%
Total	185,227	100.0%	58,529	100.0%

Source: Fiat Annual Report, 2007

comprehends educational facilities such as a course in industrial design at the engineering university Politechnico and the courses offered by the Turin branch of the *Istituto Europeo di Design* (IED, European Institute of Design), founded in Milan. The industrial tourism programme 'Made in Torino' includes visits to several key actors in the automotive cluster: apart from Fiat, also New Holland Kobelco, Iveco, Giugaro, Pininfarina, Bertone and IED can be visited. In this section, however, our analysis is restricted to the Fiat Group, and the Automobiles unit in particular.

Fiat Group Automobiles has a long tradition in providing opportunities to various stakeholders (politicians, investors, professionals, media, students, consumers, and so on) to visit the company's production sites. Company visits are seen as a public relations and marketing tool, not only with the aim to promote the Fiat brands, but also to make young people enthusiastic about a career in the automotive sector. The company wants to show the audience that car factories are not the gray, dark and sad places people might have in mind. Visits are also used to present Fiat as a good corporate citizen explaining how the company deals with its social and environmental responsibilities (by showing for example the working conditions).

The public relations department organises visits for groups only (up to 48 people at one time) from Monday to Friday. Requests from individuals and travel agencies are not accepted, and with the exception of the last class of Junior High School, people under the age of 16 are not allowed. Visitors are required to provide detailed personal information such as their age and nationality, not only because of security but also to keep track of visitor statistics. In 2007 the Mirafiori Plant was visited by 23,054 people. Visitor numbers are considerably higher during spring and autumn. The visits, free of charge, last more than two hours and start at 8.45 a.m., 10.30 a.m., and 2.15 p.m. Visits are carried out by technical guides who are professionally trained by the public relations department. Many of the guides (35 in total) are students or graduates. Because they get a lot of confidential information, the company prefers to hire relatives of employees. Guides also have to sign a contract, in which they assure to treat confidential information with care. Visits can be arranged in many languages, including Italian, English, French, German, Spanish and Portuguese.

The standard programme combines a visit to the welding shop and the final assembly line of three models (Fiat Grande Punto, Fiat Idea and Lancia Musa) at the Mirafiori factory with a visit to the Mirafiori Motor Village. For the visit to the 'Body Plant' electrical vehicles are used, including one that is suitable for disabled people. Due to safety reasons, visitors have to wear protection glasses during the visit to the welding shop. Moreover, it is strictly forbidden to take pictures inside the plant. Professional visitors that are allowed to enter the Style Centre or Safety Centre – where crash tests are carried out – are even provided with patches to cover the cameras integrated in their mobile phones. Obviously, assembly lines of cars that have not yet been introduced to the press are covered. The Mirafiori Motor Village is the first so-called 'polyvalent' sales centre of the Fiat Group, accommodating showrooms for each brand in a distinctive ambience, reflecting the specific values of the various brands. This complex with 3,000 square metres exhibition space includes a merchandise area, a playground, shops, and a bar/restaurant. Outside the complex visitors can test drive one of the available cars on a specially designed area, which is part of the Village (covering 90,000 square metres). Some cars in the showrooms are equipped with a system that enables visitors to experience a virtual test drive. Fiat sees the Motor Village not only as a collection of showrooms but also as a laboratory to demonstrate and test new marketing ideas, to be transferred to regular car dealers in a later stage. Nevertheless, competition with other car dealers in the region cannot be avoided: the Motor Village attracts audience from a wider area than normal dealers do.

Since 2005, Fiat is involved in the 'Made in Torino' tour organised by Turismo Torino (and the Chamber of Commerce). In contrast with groups each visitor has to pay an entrance fee of €10 to cover the costs (like insurance, transportation and the guide) made by Turismo Torino. The maximum group size is 30 instead of 48. For Fiat the visitors that come in via Turismo Torino are less interesting from a commercial point of view: groups mainly consist of residents and tourists. Their share in the total number of visitors is fairly small, with less than one per cent. Apart from the 'Made in Torino' programme, residents and tourists can always visit the Motor Village, which is open to the public, but admittedly this complex can hardly be categorised as an industrial tourism attraction. Concerning industrial heritage, the Lingotto area is worth a visit: this former Fiat factory has been redeveloped into a modern complex with concert halls, a theatre, a convention centre, shopping arcades and a prestigious hotel. For some professional visitors, a visit to this historical site is included in the programme, but leisure tourists have to go there on their own.

Gufram

Gufram is based in the Turin region and has two parallel core businesses: the production of special design objects ('design') and the production of armchairs and related products for large public venues such as theatres, cinemas, auditoriums, conference centres, hotels and universities ('seating'). Its relatively small production site – covering a surface of 7,500 square metres – is located in a small municipality called Balangero, with just over 3,000 inhabitants, situated to the northwest of the City of Turin, relatively close to the airport. Since 2004 the company is member of the Poltrona Frau Group, a leading international manufacturer of luxury home furnishings which was founded in Turin in 1912. Within this Group Gufram is expected to expand its activities, notably in China (Shengzen), where the company opened a new facility (12,000 square metres) to produce armchairs for the Chinese market (Cheuvreux, 2007).

The company was founded in 1952, first as a producer of armchairs and sofas for the domestic and international market. However, because of fierce global competition the company had to redefine its strategy in the 1960s and 1970s. In

1966 the company begun to produce the design objects that became part of the so-called 'I Multipli' collection in 1972. This collection comprehends objects such as a coat rack in the shape of a cactus and a couch in the shape of lips. Since then various other objects have been introduced such as a chair in the shape of a golf ball on a tee. In 1978 Gufram started its activities in what they call the contract sector, furnishing large public venues, such as the Regio Theatre in Turin.

The first core business targets individuals who are interested in buying extraordinary specially designed furniture, and who are able to spend a substantial amount of money on these objects (the cactus coat rack carries a price tag of about €3,000). The second core business targets companies and public organisations that own and/or operate major venues. The number of customers is limited to about 200 worldwide. The share of the first core business (design) in the total turnover is considerably lower than the share of the second core business (seating), but the share of the design business is growing.

In 2006 the Chamber of Commerce approached Gufram to check its availability for the industrial tourism programme. In the vision of the Chamber, Gufram had to become one of the pillars of the design theme together with car designers such as Pininfarina and Giugiaro. Gufram accepted the invitation, a decision which was mainly driven by the company's first core business, the production of design objects. Since the design business targets individual consumers, it is considered more relevant for company visits than the seating business. One of the main reasons for Gufram to open its doors for visitors was to improve its visibility in the market for design objects. Another reason was to build and maintain good relations with the City of Turin and other local authorities: an argument which is more related to license-to-operate and corporate social responsibility motives.

The industrial tourism programme of Turismo Torino and the Chamber of Commerce comprehends two visits to Gufram, one in the spring and one in the autumn. Apart from these two occasions, individuals are not able to visit the company. The size of the group varies between 20 and 25, implying a total number of visitors of not more than 50 on an annual basis (professional visitors and visitors with an educational motive *not* included). The majority of the visitors (about 70 per cent) come from North Italy. Only 10 per cent of the visitors are foreigner, leaving 20 per cent for the rest of Italy. Many visitors are retired (50+). The minimum age is 16. Tours are offered in Italian, French and English. The price of a ticket is €10. The duration of a visit is half a day, including transport. A visit to Gufram commences in the centre of Turin, in front of Turismo Torino's information point, from where a coach drives the group to the factory. Because of the peripheral location of the factory this journey takes quite some time (about one hour). The visit itself takes approximately one hour, and after that visitors get the opportunity to ask questions for about half an hour. During the tour people get to know more about the technology used (which is generally quite simple), but more importantly, also about the philosophy of the company. One of the highlights of the tour is the painting process: visitors can see how one of the designer products (like the cactus coat rack) is being painted. Interestingly, the company sometimes reschedules the painting job to make the visit more attractive. This effort fits in the philosophy of the company that participation in the programme is only effective with full commitment.

Although Gufram is fully aware of the benefits of the industrial tourism programme, the added value of the two company visits should not be exaggerated. In comparison with other marketing instruments (such as the presence of Gufram at fairs and exhibitions) the impact on income (sales) and image is limited. However, this also applies to the investments that have to be made. As the tours are organised by Turismo Torino, there are only a few things left to be done by the company's marketing department. During the visit one or two technical guides accompany the guide provided by Turismo Torino and the head of the marketing department gives a presentation. Hence, it is only a matter of some management time.

One reason for the limited impact on sales is the fact that the individuals that enter the factory do not necessarily belong to the target groups of the design business. Many of the visitors would never be willing or able to buy the expensive design objects. It is a very mixed audience. For the same reason, the industrial tourism programme – at least in its current setup – is not the most suitable instrument to reach potential employees. The reception of groups from (polytechnic) universities is much more useful in this respect: professional visitors and visitors with an educational motive can contact the company's marketing department directly to get a factory tour (for free). However, despite this target group mismatch the company argues that the participation in the programme – including the visibility on the brochures and the website of Turismo Torino – helps to promote the Gufram brand. The added value could be further enhanced if Turismo Torino and the Chamber of Commerce could somehow succeed to realise a better match between the target groups of the company and the composition of the groups that visit the firm. According to Gufram this could either be done by combining a visit to Gufram with a visit to another design company on the same day or by introducing themes for every month, with design as one of the themes. Both strategies would help to attract an audience with a stronger interest in design: people who are more likely to buy Gufram's objects.

Aurora

Aurora is a producer of writing instruments, luxury leather goods and paper. The company, founded in 1919, is owned and managed by the Verona family that since four generations is in the writing business. After the company's offices in the city centre were destroyed by bombardments in the Second World War, the firm decided to relocate its headquarters to a historic building in a suburb to the north of Turin. This building is located alongside the old Stura Abbey: an historic monument comprising a church and a cloister. The production site covers an area of 10,000 square metres. Aurora employs more than 100 people. The company's products are exported to more than 50 countries around the world (notably to America, Asia and Europe); the Italian market is however still by far (55 per cent) the most relevant. In Italy the company has a relatively dense sales network with more than 1,000 shops, including 150 very exclusive ones. Important growth markets for the firm are China and India where several Aurora stores and boutiques are operated (like in Shanghai and Beijing).

The current director, Cesare Verona, has been responsible for an important change in the strategy of the company, with implications for the firm's attitude towards visitors. Under the direction of his predecessor (his father) Aurora used to be an inward-looking company, keeping outsiders out as much as possible. This strategy can be related to the anti-capitalist terror in the 1970s. Cesare Verona changed the company into a more open minded and outward looking firm, well embedded in the local environment and with an open doors policy for visitors. Two motives explain the reorientation: on the one hand the new approach fits well in a network economy, in which every contact with the outside world can add value, either directly or indirectly; on the other hand it responds to the increasing competition on a global level, forcing a company like Aurora, like many other European firms, to focus on high quality rather than low prices. The fact that Aurora is produced in Italy adds a premium to the brand and hence, to the products that carry the brand. The company considers industrial tourism – in addition to several other networking activities such as giving presentations at universities and schools – as a perfect instrument to emphasise the link between the product and the region in which it is produced.

In 2006 Aurora started to open its doors; initially for distributors only, but later also for other professional (technical) visitors including schools and governments. Today, the tours target all the company's stakeholders: customers, bankers, suppliers, employees, and so on. Between 2006 and 2008 about 4,500 visitors have been counted. Most of them are Italian, but also foreigners (for example from Germany and the US) have found their way to the site. Older, retired people (50+) are relatively well represented. During the past years the pricing strategy changed radically. In the first year the tour was offered free of charge. In the second year the company begun to charge €5 to cover insurance costs. In the third year Aurora introduced three different tour packages: a basic version (€10), an advanced version (€15) and a luxurious version (€20). These prices include €5 for the insurance. The tours are only offered to groups with a minimum size of 20 and a maximum size of 50. Group visits can be organised on any day, and at any time of the day.

In 2007 Aurora accepted the invitation of Turismo Torino and the Chamber of Commerce to participate in the industrial tourism programme. The company is considered one of the key representatives of the pen cluster, together with other producers such as Lecce Pen. In contrast with (professional) groups, individuals can visit Aurora only twice a year, via Turismo Torino. The entrance fee is €10 which has to cover insurance costs, but also transportation costs made by Turismo Torino (picking up tourists in the city centre and driving them to the factory, and back). Remarkably, the tour organised by Turismo Torino is available in Italian only, while professional visitors can be guided in other languages (at least in English) too.

The aim of the tours that are offered is not only to show the tangible products (pens), but also to create a positive feeling around the product and the company. Visitors get to learn more about the history of the company, the products that have been produced, the production process, the design process and the repair process. When they arrive, visitors first see a video that provides an overview of the company's history and activities. After that they get a full tour in which they see every element of the production process (design, production, repair) from nearby: the distance between the working staff and the visitors is remarkably short, allowing face-to-face communication. In addition, visitors get to see several photo galleries with pictures of famous people who visited the company and, last but not least, the products produced by the company.

Aurora clearly expects substantial benefits from industrial tourism. Company visits fit in the firm's strategy to adopt an outward-looking approach, resulting from a combination of marketing objectives (branding, customer satisfaction and loyalty) and license to operate motives ('being a good citizen'). The fact that benefits are expected is also reflected in the investments that have been made: information signboards (in several languages), photo galleries and even pieces of art in the workshop add to the tour's attractiveness. Interestingly, director Cesare Verona took inspiration from several other company tours around the world, for instance the one at Ferrari.

5 Applying our frame of analysis

In the third section we described the strategy and actions of the City of Turin and the Piedmont region with regard to industrial tourism, introducing the Made in Torino tour. In the fourth section we analysed the role of three companies in this tour, but also paid attention to other activities that could be categorised under the heading of industrial tourism. In the present section we apply the frame of analysis to the observations we made in both sections. We analyse the conditions under which industrial tourism can contribute to the realisation of strategic objectives of city and enterprise. The issues that we address are 1) adequate target group selection; 2) the attractiveness of the product; 3) benefits in terms of income and image for city and enterprise; and 4) the organisational setup.

Adequate target group selection

The first issue to be addressed is the selection of target groups. Here we should make a distinction between the groups that are targeted and the groups that actually visit the companies. Following the vision and strategy to improve Turin's image, and to promote the excellence of Turin-based companies, the Made in Torino tour could be expected to target nearly all relevant stakeholders, though with a strong emphasis on potential ambassadors of the city. As the programme is relatively small (in terms of visitor numbers) it needs multipliers to make a difference. In practice, however, the programme is mainly oriented towards visitors and residents with a leisure motive, which is reflected in the fact that people are picked up by a coach in the city centre. As is common for industrial tourism in which the leisure motive is dominant, older and retired people (50+) are overrepresented (with a share of more than 60 per cent), raising the question if they are the ambassadors or multipliers the city needs to change its image. Young people (under 25), that are potentially more effective as multipliers since their networks are generally larger and expanding, are underrepresented with only 5 per cent of the visits. The minimum age of 16 not only explains the low share of young people, but also the low share of people between 26 and 50, as many of them are parents of children under 16.

Visitor statistics also demonstrate that local residents (with a share of 70 per cent) are very interested in the programme and indeed act as ambassadors, by inviting visiting friends and relatives to join on a second or third visit to the same company. Although the programme wants to improve the image of the city internationally, the share of foreigners is relatively low with only 10 per cent. The recent decision to improve the availability of tickets for tourists shows that the initiators also want to reach an international audience, not only indirectly through media attention but also directly through actual visits, which is in line with the underlying vision and objectives.

The people that visit firms through the Made in Torino tour are not necessarily the target groups that participating companies want to reach. Concerning the scale of the target groups the differences between the industrial tourism programme and the strategies of the companies are considerably small. The companies that we analysed operate on a global level, but with a more than average presence on the Italian market, which is to some extent in line with the composition of the groups delivered by Turismo Torino. Differences do appear in the scope of target groups. In practice companies adopt different approaches with regard to the people they want to receive depending on the benefits they expect from industrial tourism. The three firms we analysed seem to prefer visitors with a business or educational motive above people with a leisure motive. Apart from the Made in Torino tour they open their doors for various guests, but in general, only for groups (not for individuals) with a clear stake in the company (like suppliers, customers and potential employees) or actors that are relevant for the license to operate (for example media and politicians).

For Fiat, producer of consumer products, this implies that nearly every individual can visit the factory all year long – as potential buyer of a car, shareholder or employee – as long as you find a group to join you. This can be any association or club; Fiat is not as choosy as other companies are. Gufram is an example of a company that is more selective in this respect. Apart from the Made in Torino tour the company only opens its doors for professional visitors and students. If they were only in the contract business, they would probably not participate. But also for the design business it has to be admitted that many tourists and residents that come in are not potential buyers of the design objects. For Gufram the groups delivered by Turismo Torino are too diverse from a sales point of view, implying that other motives should explain the company's involvement. Aurora clearly takes a broader view: this company has expanded the target group scope of the company visits gradually from distributors only to all stakeholders. The director believes that any contact with the outside world generates added value, as visitors share their experience with friends, relatives and business partners.

The attractiveness of the industrial tourism product

The second issue is the attractiveness of the industrial tourism product which is partly dependent on the accessibility. Turismo Torino, the Chamber of Commerce and the individual companies have made quite some efforts to make the industrial tourism product more attractive. The products are easily accessible through the Turismo Torino website, which currently provides information in Italian, English and French, but which will be expanded by sections in Spanish and German. Also brochures are available in several languages. Tours can also be booked at one of the three strategically located information offices. For tourists the accessibility of the sites is excellent, with buses leaving from the city centre. By showing a video about the company or by giving a presentation, the bus tour – which can take up to one hour – has been made more attractive. Only sites that can be reached within a one hour drive can be included in the programme. The idea behind this is that a visit including travel time should not last longer than one morning or one afternoon (four to five hours).

Groups that participate in the Made in Torino tour are guided by hostesses of Turismo Torino. The maximum size of a group is 25, for safety reasons but also to keep the tour attractive. During the visit to a company they are often accompanied by representatives of the firm, often the same people who take care of the professional visitors and students, or other groups that find their way to the company directly. It is our experience that it is often more interesting to be guided by a manager of the firm (which is common at Gufram and Aurora) rather than by 'professional guides' hired by a company (Fiat's policy). At Fiat we noticed that the tour was not as authentic, interactive and tailor made as at Gufram, or to make the comparison with a competitor, as at Volkswagen where they recruit guides among the employees on the work floor. This had nothing to do with the quality of the guide who had a lot of knowledge and was very experienced.

The three companies that we analysed in detail all deliver an attractive industrial tourism product. Fiat has the most comfortable position as cars are appealing to many people. Moreover car plants – like the Mirafiori factory – use a logistics process which is very appropriate for the reception of visitors. Fiat has invested in electronic cars with audio equipment to show visitors the assembly line, and some other parts of the production process. Furthermore the entrance hall of the factory has recently been refurbished, giving the visitor a warm welcome. Another strategy used by Fiat to make the product more attractive is to combine the factory visit with a visit to the Motor Village, and for some groups, to the historical

Lingotto site. Gufram is particularly interesting for people who are familiar with the design objects: the building itself is not as spectacular as the Fiat factory, but it is an experience to see and touch the objects, and to see how they are produced. The company does everything to make the visit more attractive, even resulting in changes in the planning of the production process. Aurora has the advantage of being located in a very attractive and historical building. A visit to the company is attractive because it provides insight into every aspect of the production process; moreover, the distance between the visitor and the workers is very short. The company has invested in a video, art objects, a photo gallery and information displays to make the visit more attractive.

Benefits in terms of income and image for city and enterprise

For the companies we analysed company visits produce several benefits. Fiat, Gufram and Aurora consider company visits as important instruments of their public relations and marketing strategy. By opening their doors for students, (potential) business partners, media and politicians they want to improve their reputation, which in the end should lead to a better financial-economic performance. The direct income from company visits is very limited. The Fiat tours are free of charge, and also Gufram doesn't charge any money. The only exception is Aurora which has discovered company tours as a modest source of income, with entrance fees ranging from €10 to €25. The benefits of the Made in Torino tour, however, are less obvious. For the companies there is no direct income. For Fiat, tourists and residents are less interesting from a commercial point of view, and the number of people that come in through Turismo Torino is only a fraction of the total number of visitors (less than 1 per cent). Also for Aurora the share of Made in Torino visitors in the total number of visitors is small (approximately 5 per cent).

Aurora and Gufram have in common that the visitors brought in by Turismo Torino are not necessarily potential buyers of their relatively expensive products. Nevertheless both companies expect indirect benefits, not only through word of mouth advertising but also through media attention for the programme and the promotional efforts of Turismo Torino and the Chamber of Commerce (by websites, brochures, and so on). Aurora considers participation in the industrial tourism programme as one of the many actions to strengthen the link between the company and its location. The fact that Aurora pens are produced in Italy is very important in the marketing strategy, and enables the company to sell pens in the higher price segments of the market. This also applies to Gufram's design objects. Although Fiat has not explicitly mentioned the relation with Turin as an argument to be involved in industrial tourism, it cannot be neglected that the company and the city are strongly interconnected: they obviously need each other. Gufram admits that the relation with the City of Turin is probably the primary reason to be involved in the programme: the company wants to be a good citizen.

Fiat, Gufram and Aurora are consumer-oriented companies. Their involvement partly depends on marketing and sales motives. However, if we look at the list of companies that participate in the Made in Torino tour we have to conclude that also business-to-business companies are involved. The involvement of Iveco and New Holland Kobelco might be related to the involvement of Fiat, but for the aerospace companies and car design companies other benefits must play a role. Probably they have an interest in developing the local aerospace and local design cluster respectively, securing access to the labour supply, and keeping good relations with the city and its inhabitants.

The direct benefits of the Made in Torino programme for the city are small. With a maximum number of participants of 1,200 (40 visits × 30 participants), the economic impact is moderate. All direct income from ticket sales is needed to cover the costs, and still then, the Chamber of Commerce needs to close the financial gap that remains. Dividing the €100,000 subsidy by 1,200 participants we can conclude that each visitor is subsidised with €83.33 (or to put it otherwise: the actual costs for one visitor are €93.33). This amount does not even include the labour costs made by the Chamber of Commerce. Nevertheless, the Chamber of Commerce and Turismo Torino are very positive about the results of the programme so far. Probably this is because the programme resulted in a lot of attention from the media and the academic world (including *our* attention!). As a consequence many people have heard about the Made in Torino tour without having participated. Made in Torino is more than a tourist product and it is more than a tourism campaign: it fits in a broader vision and strategy to promote the city as a tourist destination and business location, with the Olympic Winter Games 2006 as a catalyst. Compared with other city branding campaigns, the €100,000 subsidy of the Chamber of Commerce is modest, but not less effective. It results in positive media attention, and it makes people feel proud of their city and its centres of excellence. Although this has not been confirmed during our research, the tour could also stimulate companies to open their doors more often, though not necessarily via Turismo Torino.

The City of Turin and the companies have an interest in industrial tourism, but does that also imply that the interests of city and enterprise are converging? If we see the Chamber of Commerce as a representative of the private sector, the answer would be yes: the Chamber supports the programme because it believes that the performance of the city and the companies located there are interrelated. Both city and enterprise have an interest in displaying a more realistic, more positive picture of the manufacturing industry in Turin. The three companies that we analysed also seem aware of their relation with the urban environment in which they operate. For Fiat good access to the labour market is a very important issue. In the cases of Gufram and Aurora the location is a key asset in the marketing strategy. Nevertheless, the involvement of the companies in the Made in Torino programme could be further improved: currently the companies get free exposure, and there is no willingness among the firms to contribute to the programme financially (apart from investments in the site and some management time). Other companies refused to participate or ended their participation (like pen producer Stilolinea), mostly because they are more concerned about their core business, and the increasing competition on a global scale. Industrial tourism is by many considered a side activity that only costs money, while only a few – with Aurora as one of the best examples in our sample – see it as an opportunity strongly related to its core business.

For both the city and the region, industrial tourism is a niche market. The total number of visitors is for sure considerably higher than the 1,200 participants of the Made in Torino tour, but still their share in the bed night figures is probably very low. While Turismo Torino presents industrial tourism as one of the key assets, the region sees it as a small insignificant market. This is understandable particularly in view of the fact that the distribution of national funds to regional tourism organisations fully depends on bed night statistics, and not on foreign direct investment (new business development). Moreover the city and the region already have a rich tourism offer with the mountains, the lakes, the hills and the city of Turin itself.

The organisational setup

In the three companies we analysed company visits are fully integrated in the marketing and/or PR departments: industrial tourism is considered a tool of reputation management. Furthermore we can observe strong involvement of managing and marketing directors, notably in the smaller companies (Gufram and Aurora), where they often show visitors around themselves. Not surprisingly, Fiat has a more professional organisation with a relatively large PR department and about 35 guides (students that are paid for every hour they work).

On the level of the city, or actually the Province of Turin, the tourism organisation Turismo Torino and the Chamber of Commerce have together developed an organisational environment for industrial tourism. Their cooperation confirms that industrial tourism requires cooperation between the fields of economics and tourism. Both organisations take an active role in the partnership: Turismo Torino takes care of developing promotional materials and organising the tours, while the Chamber of Commerce acts as intermediary between Turismo Torino and the participating companies. Both organisations take care of developing and implementing the strategy, resulting in the selection of clusters and companies. They both act in line with the Strategic Plan 2000–2010 that was developed in consultation with all relevant stakeholders, trying to improve the image of the city, creating a better business climate and attracting more tourists. The regional tourism organisations – the regional board for tourism, trade and sports and the research and development agency – do support the initiative (they are not against it), though not financially.

An important organisational element of Turin's industrial tourism programme, one that could be interesting for other cities to learn from, is the step-by-step expansion. Every year one or two clusters are added to the programme. This strategy has several advantages: 1) the costs of approaching new participating firms and developing new tours are spread out over several years; 2) the programme can be started quickly resulting in positive media attention that may stimulate other companies to participate too; 3) the addition of a new cluster can be used to attract more media attention and, related to this, 4) the industrial tourism product is continuously improving.

To conclude, the fact that the Chamber of Commerce finances the Made in Torino Tour completely raises the question if companies actually recognise their interest in the programme. On the hand, we found that the target groups companies want to reach (customers, potential employees, decision makers, etc.) are often not well represented in the groups brought in by Turismo Torino. In general, the groups include many old people (50+) who are often considered less interesting not only from a human resources perspective but also from a marketing and public relations point of view. Companies are generally more interested in receiving primary stakeholders (technical visits), which they take care of themselves. According to one discussion partner, the Made in Torino Tour could become more beneficial for companies if the initiators somehow succeed to attract more specific target groups, for instance by developing thematic packages. On the other hand, we have observed that companies do participate in the programme – mainly through investments in management time – because they want to be good citizens and they understand that relations with society in general, and the local authorities in particular, are important.

References

Berg, L. van den, Braun, E., and Otgaar, A.H.J. 2002. *Sports and City Marketing in European Cities*. Aldershot: Ashgate.

Chamber of Commerce for the Turin Province. 2008. [Online]. Available at: http://www.to.camcom.it [accessed: 2008].

Cheuvreux 2007. *Poltrona Frau* [Online]. Available at: http://www.borsaitaliana.it/mediasource/star/db/pdf/62875.pdf [accessed: 2008].

City of Turin 2000. *The strategic plan of Torino 2000-2010.* [Online]. Available at: http://www.recs.it/en/upload_pdf/The_Strategic_Plan_of_%20Torino.pdf [accessed: 2008].

City of Turin 2005. *NumeraTO: Counting Torino.* [Online]. Available at: http://www.comune.torino.it/statistica/osservatorio/numerato/en/index.htm [accessed: 2008].

European Cities Tourism 2007. *ECM-European Cities Tourism Report 2006/7; European city destinations take it all*, Press release from European Cities Marketing, January 15th 2007. [Online]. Available at: http://europeancitiestourism.episerverhotell.net/upload/Press%20release%202006-07.pdf [accessed: 2008].

Invest in Torino Piemonte 2008. http://www.investintorinopiemonte.org.

IRES 2007. *Piemonte Economico e Sociale 2006: I dati e i commenti sulla regione.* [Online]. Available at: http://www.regiotrend.piemonte.it [accessed: 2008].

ISTAT 2007. Available at: http://www.istat.it [accessed: 2008].

Maggi, M. and Piperno, S. 1999. *Turin: The vain search for gargantua*, Working paper no 124. Torino: IRES Piemonte.

Piedmont Region n.d. *Tourism in Piedmont: the figures*. [Online]. Available at: http://www.regione.piemonte.it/lingue/english/pagine/cultura/approfondimenti/02_piemontur_en.pdf.

Rosso, E. 2004. *Torino: Policies and actions at a metropolitan level*. Paper to the conference: La Gouvernance Metropolitaine: Recherche de coherence dans la compléxité, Montréal, 7–8 October 2004. [Online]. Available at: http://ejc.inrs-ucs.uquebec.ca/Torino.pdf [accessed: 2008].

SAGAT 2006. *Turin Airport Statistics 2006*. [Online]. Available at: http://www.turin-airport.com/pdf/statistiche_2006_eng.pdf [accessed: 2008].

Sviluppo Piemonte Turismo 2007. *Dati statistici sul Turismo in Piemonte*. [Online]. Available at: http://www.regione.piemonte.it/turismo/osservatorio/rapporto2007.htm [accessed: 2008].

Symcox, A.L. and Symcox G.W. 2006. *A history of Turin*. Torino: Einaudi.

Winkler, A. 2007. *Torino City Report*. Torino: Centre for Analysis of Social Exclusion. [Online]. Available at: http://sticerd.lse.ac.uk/dps/case/cr/CASEreport41.pdf [accessed: 2008].

Discussion partners

Marzia Baracchino, Dirigente Settore Promozione Turistica, Direzione Turismo, Commercio e Sport Settore Coordinamento della Promozione domanda Turistica – Organizzazione degli Eventi Promozionali.

Cristina Cerutti, Area Prodotto – City Break, Turismo Torino e Provincia.

Fabrizio Gardella, Amministratore Delegato Direzione Commerciale/ Marketing, Gufram.

Gianpiero Masera, Chief Executive, Promotion and Territorial Development, Camera Di Commercio Industria Artigianato E Agricoltura Di Torino.

Gabriella Ragazzone, Communications Public Relations, FIAT Group.

Maria Elena Rossi, Chief Executive Officer of Sviluppo Piemonte Turismo.

Cesare Verona, Managing Director, Aurora.

Chapter 6
Industrial Tourism in Shanghai

1 Introduction

This case study comprehends a description and analysis of industrial tourism in the Chinese metropolis of Shanghai. There are several reasons why we selected Shanghai for this comparative research. Firstly, Shanghai is the largest economic, financial, trade and logistic centre of China and is also commonly known as 'factory of the world': the city possesses substantial amounts of industries in many different sectors. Secondly, the public influence is a lot stronger in China, when compared to European countries, and thus we can expect that also the public programmes for support are rather mature in Shanghai. And indeed, public initiatives to promote industrial tourism in Shanghai have already started in 1997.

The structure of this case study follows the subsequent pattern: after this introduction, section 2 deals with the characteristics of Shanghai, including a description of the city's tourism profile. In section 3 we provide an overview of the way industrial tourism is organised, after which section 4 analyses some industrial tourism products in Shanghai's metropolitan region. Next, section 5 confronts the Shanghai case with our frame of analysis and draws several conclusions.

2 General profile of Shanghai

Shanghai is located on the banks of the Yangtze River Delta and on the shore of the East China Sea and boasted a population of more than 18 million inhabitants in 2006. The city has been growing steadily over the past years: since the year 2000 it gained more than 2 million inhabitants (Statistical Yearbook, 2007). Shanghai covers an area of over 6.300 km^2 (Shanghai Statistics Bureau, 2008) and is one of the four city-provinces in China. Shanghai is a province (comprising 18 districts and 1 county) and a municipality at the same time, and legally speaking it is only one tier below the State.

Shanghai is well endowed with transport infrastructure: there are two international airports with a total annual passenger transport volume of 30 million persons. These two airports will be connected by a high speed train line (Maglev, over 430 km/hr) in 2010. One high speed Maglev-line is already in operation and connects Pudong International Airport with the centre of Shanghai; this line is the only commercial Maglev-train line in the world. Moreover, there are two railway stations which dispatch over 100 trains back and forth from Shanghai every day. There are several freeways that connect Shanghai with other cities. By the end of 2010, there will be 12 subways in Shanghai. Notwithstanding state of the art transport infrastructure, traffic congestion is still a big challenge for Shanghai. During the rush hour, spending one hour for covering 4 km by car is a normal phenomenon.

Political and economic context

Shanghai is the cradle of Chinese industry: already 400 years ago the city was a centre of fishing, trading, textile and finance industries. In 1990 the share of the secondary sector in Shanghai's GDP was 63.8 per cent. Since then, the share of industry declined somewhat, as the city's economy gradually became more service-based. In 2005, the share of industry was 48.9 per cent, which is lower than 15 years before, but higher than in all major European cities.

In order to understand the economy in Shanghai and the Chinese economy in general, some political-historical background is needed. Many changes in the Chinese economy can be related to the year 1978 when Deng Xiaoping, China's leader at that time, introduced a number of reforms that were intended to promote foreign trade and economic investment. This so called 'open door policy' brought drastic changes in communist China, as before these reforms (from 1949 until 1978) only public companies were allowed to operate in China.

Since 1978, two other ownership models for companies are legally possible, apart from public ownership. First, both Chinese citizens and foreigners are allowed to operate businesses that are 100 per cent privately-owned. A second legal entity is the public-private joint-venture company that is jointly owned by a public Chinese body and by a private foreign investor. To sum up, nowadays companies can take four different legal forms in China: 1) a public company, 2) a public-private joint venture, 3) a foreign-owned private company and 4) a Chinese owned private company. In 2007, 8.34 per cent of all companies in Shanghai were state owned, 32.19 per cent were collectively owned (joint venture), 23.91 per cent were foreign-owned private companies, and 35.58 per cent is privately owned (Statistical Yearbook, 2007).

The city province of Shanghai has maintained a GDP growth rate of over 10 per cent over the past 11 years. Economically speaking, Shanghai performs far

Table 6.1 Shanghai's industrial structure: composition of the Gross Domestic Product (GDP)

Year	Primary Sector	Secondary Sector	Tertiary Sector
1990	4.3 %	63.8 %	31.9 %
1995	2.5 %	57.3 %	40.2 %
2000	1.8 %	47.6 %	50.6 %
2005	0.9 %	48.9 %	50.2 %

Source: Shanghai Statistics Bureau, 2008

above the Chinese average. The GDP per capita of China in 2006 was $1,553.87 while Shanghai's level was $7,189 (Econstas, 2006). While Shanghai covers only 0.06 per cent of the land of the Chinese state and accounts for only 1 per cent of the population, its share in the Chinese GDP, financial income, and import/export revenue is considerably larger: 4.90 per cent, 12.2 per cent and 24.4 per cent respectively in 2007 (Statistical Yearbook, 2007).

Tourism profile

Tourism in Shanghai before 1978, the year of the launch of national open-door policy, was characterised by group package tours, with little emphasis on attracting tourists on an individual or independent basis. Since 1978, Shanghai's tourism policy has been trying to extend this focus. The number of visitors had been increasing over the past years and in 2007 the city received a total number of 96.84 million visitors, of which 23.57 million were overseas tourists, 73.27 million domestic tourists and 6.05 million local tourists (Statistical Yearbook, 2007). So clearly, domestic Chinese visitors are the most important tourist group in Shanghai. Domestic tourism in China has experienced a huge increase over the past few years, and the number of domestic tourists more than doubled between 2003 and 2007 (China National Tourism Statistics, 2008). The most important overseas visitors in Shanghai are the Japanese, followed by Chinese citizens living abroad.

The increasing number of tourists went along with an increasing importance of the tourism sector for Shanghai's economy. In 2004 the tourism sector accounted for 6.4 per cent of the city's total GDP; this number rose to 7.6 per cent in 2005 (Statistical Yearbook, 2007). By the end of 2007, 320 hotels, of which 32 five-star hotels, were in operation in Shanghai. Three- and two-star hotels are the main hotel types with a proportion of 33.75 per cent and 45.4 per cent respectively.

Tourism attractions Shanghai offers big city dynamics and a wide array of amenities in the fields of sports, leisure, culture, nightlife, and so on. The most famous tourist area is the so called Bund, which is an attractive waterfront area

Table 6.2 The origin of overseas visitors of Shanghai

Position	Country	Visitor number
1.	Japan	1,369,000
2.	China	1,203,000
3.	Taiwan	725,000
4.	USA	546,000
5.	Hong Kong and Macao	477,000
6.	Germany	218,000

Source: Statistical Yearbook, 2007

along Huangpu River. This area had initially been a British settlement, and later one became the 'International Settlement' of Shanghai. Today this area still boasts an important agglomeration of historic European style buildings. Apart from this prime tourist area, other attractions include the famous shopping district 'Nanjing Road' and the 'Fake market', where imitations of many branded goods can be purchased. Shopping as such is one of the prime tourist attractions of the city. Shanghai also hosts a number of museums, of which the Shanghai Museum of Art and History is the most well known one. Moreover, Shanghai offers all different kinds of architectural attractions: the remaining traditional Shanghai style constructions (temples, dwellings) and modern installations represent an assemblage of different styles of architecture. Architecturally speaking, the skyscrapers of the Pudong area (facing the Bund area, on the other side of Huangpu River) draw most attention. The skyline of Pudong, with the characteristic Oriental Pearl Tower and the Jin Mao Tower, has already become a sort of landmark for Shanghai. Moreover, Shanghai is famous for events ranging from international festivals, exhibitions and conventions to mega-sport games. The forthcoming World Expo 2010 is only one of these events, other important events include the Shanghai Formula 1 race and a Tennis Grand Slam tournament.

Most attractive areas and places of touristic interest in Shanghai are concentrated in the centre and thus, also leisure tourists concentrate in these areas. The public transportation system is a very convenient mode to reach the tourist attractions in or near the city centre. However, for some of the newly developed tourist attractions in the suburbs public transport connections are still missing or are rather inconvenient. Although Shanghai does have a range of interesting tourism facilities and attractions, it boasts less places of prime tourist interest than other cities of similar size – for example, the Chinese capital Beijing has a lot more sites of historic interest. Nevertheless, Shanghai receives more visitors a year than Beijing (according to the China National Tourism Administration, 2008). Due to its relative lack of natural and historic attractions Shanghai can actually be described rather as a business city than as a traditional tourist city. However, one kind of the leisure visitors is well represented, namely the shopping tourists: Shanghai is widely known as a shopping paradise and attracts shopping tourists from near and far.

3 The organisational framework for industrial tourism in Shanghai

This section presents the organisational framework for industrial tourism in Shanghai. First we discuss Shanghai's tourism policy, followed by a sub-section about the Shanghai Industrial Tourism Promotion Centre (SITPC).

Tourism policy

The Shanghai Municipal Tourism Administrative Commission (hereafter abbreviated as the Tourism Commission) was set up in October 1978. It is a government body which holds the responsibility to manage and coordinate all tourism related issues in Shanghai. This commission has much influence, but politically it has a lower standing than several other commissions such as the economic commission, the city development commission, the city construction commission or the city transportation commission. Overlapping responsibilities with other commissions and complex hierarchies sometimes lead to coordination problems between the different bodies.

In 1997, in response to the city's shortage of traditional tourist attractions, the Tourism Commission set up a milestone-like tourism development strategy to diversify the city's tourism products. Ten key urban tourism products were identified, eight of which were considered 'new types of tourism' in Shanghai. Industrial tourism was recognised as one of these new types of tourism. The strategy also defined citizens, together with business and leisure tourists, as the main target groups of the city's tourism policy. The development strategy aimed to strengthen the relation and the cooperation between the ten key urban tourist products and the involved stakeholders.

After the Shanghai municipal government had officially accepted industrial tourism in 1997 as one of the ten key sub-segments of urban tourism, policy attention to industrial tourism initially remained relatively low compared to the other key sub-segments such as leisure tourism, event tourism and MICE (meetings, incentives, conferences, events) tourism. The available information about industrial attractions and services in Shanghai was fragmented and incomplete, and only a small amount of companies organised plant tours.

This changed, however, in 2002 when the national government developed an industrial tourism policy. Every year since then the Chinese National Tourism Commission nominates a number of industrial tourism sites as 'National Models of Excellence'. Companies with this status receive subsidy from the state, and can get additional financial support from the local Tourism Commission. Moreover, these companies get free publicity. By the end of 2007, about 270 companies across the nation were recognised as industrial tourism examples for other firms, including 15 companies in Shanghai.

The new national industrial tourism policy was quickly adopted by local policy makers (as is very common in the Chinese context). In 2004 a delegation of the Shanghai government visited a number of places in Germany and France to learn more about industrial tourism and industrial heritage. This trip helped to develop innovative ideas and to shape the city's industrial tourism policy. This policy, documented in the 'Shanghai Industrial Tourism Development Plan 2006–2010', is the first local industrial tourism development plan in China. It provides a geographic and thematic overview of Shanghai's industrial tourism resources, with seven main areas and corridors of industrial tourism development. The zoning of these development areas is largely similar to the layout of the city's most important industrial clusters (Shanghai Institute of Tourism, 2007). The city considers industrial tourism not only as an instrument to improve Shanghai's image, but also to make the development areas more attractive and lively by realising a

Table 6.3 National Models of Excellence in Shanghai, 2007 (SITPC, 2007a)

Year	Number of Nominated Companies in Shanghai	Nominated Companies
2004	1	Baosteel Group
2005	2	Bridge 8 Creative Industrial Park; Shanghai Volkswagen Co. Ltd.
2006	4	M50 Creative Park; Geological Museum Shanghai; Industrial Park of Metres-Bonwe Group; Industrial Park of Tobacco Group Shanghai
2007	8	Caohejing Hightech Park; 1933 Old Factory; Chedun Film Shanghai; Shenxian Alcohol Factory; Yakult Shanghai; Dairy 8 factory; Cobtt Healthy Drinks; Space 188 Creative Park

Source: SITPC, 2007a

mix of functions and additional sources of income. For instance, in the central area, tourists can be valuable for artists and small businesses located in former industrial buildings. The plan, written in cooperation with local universities, resulted in the creation of a new entity: the Shanghai Industrial Tourism Promotion Centre (SITPC).

The Shanghai Industrial Tourism Promotion Centre (SITPC)

In 2005, in response to the new national and local industrial tourism policies, the Shanghai Municipal Economic Committee and the Tourism Commission jointly set up a platform organisation named Shanghai Industrial Tourism Promotion Centre (SITPC). The Centre, funded by the two founding organisations and employing around 30 employees, has the mission to develop a more integrated industrial tourism product, and to improve the marketing and promotion of this product. Its work area covers the entire municipality (the city province of Shanghai), which is larger than the metropolitan area of Shanghai (SITPC, 2007a). SITPC is as an independent non-profit company; it also owns, however, a profit-oriented travel agency (see further on). The Centre cooperates with numerous stakeholders of industrial tourism: companies, travel agencies, citizens, associations and government commissions. Representatives from these five stakeholder groups constitute the executive board of the SITPC. Also representatives from the two

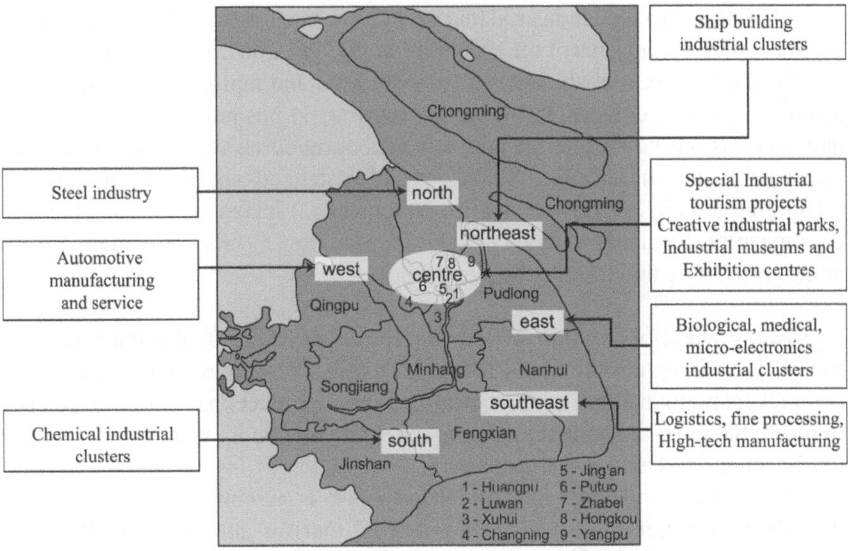

Figure 6.1 Spatial layout of Shanghai's industrial tourism resources

Source: Shanghai Institute of Tourism, 2007, translated and adapted by the authors

founding organisations are represented on the board, and are in fact the two key leading actors. The composition of the board gives an indication that SITPC has a deep relationship with the local government. Also the vision of the organisation clearly indicates the public mandate: the centre is intended to facilitate governmental strategies, to integrate and promote Shanghai's industrial tourism resources (as a package) and to build a network of key persons in the realm of industrial tourism.

There is a clear split of functions between SITPC and the travel agencies. SITPC, with its location in the city centre and its website, serves as a central contact point for people interested in industrial tourism. The organisation of visits (like the programme of the visit, the provision of a shuttle bus and the arrangement of the appointment) is, however, not a specific task of the centre. Interested parties are referred to one of the 28 travel agencies in Shanghai that have signed strategic cooperations and promotion contracts with the centre. Only agencies that meet a certain quality standard can cooperate with SITPC. To check if the agencies fulfil the SITPC standard, they are evaluated by SIPTC on a regular basis. One of the 28 travel agencies is owned by SITPC itself, which is why tours can also be booked directly at the office of SITPC in the centre of Shanghai. All agencies work on a profit basis, which means that they charge for the bus transport, the meals that they offer and so on. Apart from these costs, visitors additionally have to pay for those industrial attractions that charge an entrance fee (SITPC Sample Contract, 2007).

SITPC has classified industrial tourists into six categories, which also represent the six main target groups of the organisation (SITPC, 2007a): 1) students: pupils, middle and high school students, college students and master/MBA students; 2) nostalgic visitors: citizens and domestic tourists; 3) business visitors: domestic and overseas business professionals; 4) neighbourhood visitors; 5) young visitors (teenagers) and 6) families. SITPC has also defined six main missions: 1) integrating the city's industrial tourism resources, 2) increasing the involvement of companies, 3) quality assurance, 4) local promotion, 5) international promotion and 6) consulting and training for companies.

Integrating the city's industrial tourism resources One of the main aims of SITPC is to integrate the city's industrial tourism resources, thus creating one comprehensive industrial tourism product. Three products have been developed to reach this goal: the annual industrial tourism ticket book, visitor packages and tourist routes, and a data-sharing system.

The Annual Industrial Tourism Ticket Book is essentially a voucher-book that allows the free-of-cost visit of industrial tourism attractions in Shanghai. It was issued for the first time in 2007, with 93 vouchers, and has been further extended in 2008, providing access to 118 attractions, 19 of which are located in two neighbouring provinces of Shanghai in the Yangtze River Delta[1]. The main target group is the Chinese market, as only 23 attractions provide tours in English and are thus also accessible for international tourists (SITPC 2007a and 2007b). The Annual Ticket Book is promoted by SITPC as an incentive programme for employees of local companies. Indeed, most of them are purchased by domestic firms who give them away as presents to their employees.

Another strategy to integrate the city's tourism resources is the development of visitor packages and tourist routes. SITPC and a number of cooperating travel agencies offer several of such packages. The elements of a package are not necessarily linked to another by a theme; proximity is the main connecting mechanism. Many tours also include attractions or activities that are not related to industry. Instead the combination with more traditional attractions should increase the acceptance of industrial tourism and should help to bring more visitors in touch with the issue. Moreover, due to spatial restrictions it would be very difficult to design visitor packages from a thematic point of view: it is simply inconvenient to combine two companies in a similar field that are located on the opposite side of the city. Thus, most package tours take place within one of the seven industrial tourism zones that have been identified in the zoning plan for industrial tourism. For the design of these specific industrial tourist routes, SITPC cooperates closely with several organisations, such as the Shanghai Labour's Union, the Shanghai Youth Union, the Shanghai Female Union, the Senior Union, the Student Science

1 There are two provinces, besides Shanghai, located in the Yangtze River Delta region. These are Jiangsu province (the northern neighbour of Shanghai) and Zhejiang province (the southern neighbour of Shanghai).

& Technology Education Centre, the Light Industry Information Centre, the Agriculture Industry Association, the Service Industry Association as well as with local universities.

The third product SITPC developed in view of integration is a data-sharing system for companies that participate in industrial tourism. Already more than 100 companies have joined this network and agreed to update their industrial tourism profile on a monthly basis. In return, they can access the data of other member companies. For the year 2008, also 20 companies that are located in the Yangtze River Delta (outside the province of Shanghai) have agreed to participate in this data-sharing system.

Increasing the involvement of companies The second mission of SITPC is to increase the participation and involvement of companies. First and foremost, the centre tries to involve companies that have already opened their doors to the public. Secondly, firms that do not yet offer company tours are approached in order to persuade them to open their doors for visitors and become members of the organisation. The centre has created a database of companies and sites that could be potentially interesting partners. Selection criteria are firm size (with a preference for large companies) and the relation of a company with the region (with a preference for companies that represent the regional identity). The experience of SITPC shows that companies that meet these criteria are easier to involve than companies that don't; this particularly applies to business-to-business companies. The Centre does not only approach manufacturing firms, but also service companies: participation of service companies is considered to be crucial, as Shanghai wants to be perceived not only as a place where manufacturing takes place, but also as a creative place. SITPC employs several incentives to involve companies and to keep companies 'under their umbrella': the participating companies get free promotion by the means of newspaper supplements and can also participate for free in certain events and exhibitions.

Quality assurance In view of SITPC's third mission a standardised quality system for industrial tourism has been developed. Companies that want to become member of the centre need to fulfil several requirements. There are three main requirements that a firm has to meet if it wants to offer a company tour under the umbrella of the SITPC. First, a company needs to have the relevant 'industrial tourism hardware' such as shopping facilities for visitors, restaurants, visitor toilets, internal roads dedicated to tourists, parking facilities, and so on. The other two requirements refer to the 'industrial tourism software'. There are regulations concerning the tour guides (like sufficient professionalism and language skills and use of microphones in certain situations) and concerning the management of industrial tourism, which refers to organisational issues. Officially, SITPC is only responsible for assisting companies with the implementation of the quality standards. Eleven registered officers from the municipal government evaluate if the companies actually fulfil the formulated quality standards. These officers also check if the companies fulfil the legal regulations related to industrial tourism, and if they possess all necessary permissions.

Local promotion The fourth mission is to promote industrial tourism locally, the main tools being the organisation's website, the weekly newspaper supplements in domestic media, as well as promotional brochures and digital media (like DVDs and cd-roms). The promotional material is distributed through the 28 cooperating travel agencies, the 18 tourist offices in the city of Shanghai and through most five star hotels in the city. Moreover, SITPC participates in several events and exhibitions. The main communication channel, however, is the website (www.itripsh.gov.cn) where tours can be booked online. For students travel, agencies have special conditions: shuttle buses and all other logistic issues are often arranged for free or at a price that only covers the cost of the travel agency (thus, without delivering any profit). In order to compensate for this loss of profits, they receive a subsidy from the Shanghai City Committee, expecting that industrial tourism helps to align their studies in a better way with the demands of companies. Also other citizens have access to 'special conditions': SITPC regularly organises industrial tourism routes that are free of charge.

International promotion SITPC also aims to promote industrial tourism internationally. Although the website currently only provides information in Chinese, other promotional activities are also carried out in English and Japanese. International promotion mainly targets business visitors. Because foreign business visitors usually spend the night in accommodations of the higher categories, SITPC brochures can be found in most five star hotels in the city. For the international audience the main promotional tools are exhibitions and events. SITPC also opened its own promotional representative office in Japan, which is jointly operated with one of the biggest Japanese travel agencies (JTB Japan). The decision to open office in Japan was based on a marketing survey, which indicated that most business visitors in Shanghai are Japanese nationals. The representative office in Japan is intended to tap into this pool of Japanese business travellers, and intensify their interest for industrial tourism and for making investments in Shanghai.

Consulting and training for companies To further improve the quality of the product, SITPC assists companies as a consultant, helping them to design company tours and to build hard and soft infrastructures. Training programmes deal with 'planning and managing industrial tourism spots' covering topics such as planning, marketing, management and construction for managers and practitioners in the field of industrial tourism. The programs are intended to improve the service standard of companies involved in industrial tourism. SITPC also regularly organises meetings for its members to exchange experiences.

How successful is SITPC? SITPC has been very successful in raising the number of industrial tourism visitors: from 2 million (estimation) in the year 2004 to more than 6.15 million industrial tourists in 2006. Around 70 per cent of these visitors are pupils, students, pensioners and leisure visitors. Interestingly most of these visitors come in groups: the share of single leisure visitors is very small. The remaining 30 per cent are professional visitors. Around 75 per cent of all visitors are locals and Chinese citizens; most foreigners come from Japan or from other Asian countries. In 2006, the total revenues generated by industrial tourism were estimated at about €286,000 (all of the numbers above are rough calculations based on the data gathered from companies, travel agencies and surveys; SITPC). The number of industrial tourists amounts to 6.4 per cent of the total annual tourists the city of Shanghai received in 2006 (SITPC, 2007a), so industrial tourism is still a rather small segment of Shanghai's overall tourism economy.

To compare the results of SITPC with similar initiatives in other cities, it is important to emphasise that the centre's definition of industrial tourism is broader than our definition as it also contains industrial heritage attractions. Actually, only 18 of the 93 participating industrial tourism attractions listed in the annual ticket book of 2007 were in the category of 'individual factories'. The industrial tourism programme also provides access to modern industry parks, creative industry clusters/parks, industry museums and exhibition centres, and industrial urban landmarks. With 33 attractions, museums and exhibition centres even account for the largest share. In 2007 the most popular industrial tourism sites in Shanghai were creative parks and factory visits. Industrial museums and exhibition centres are gaining popularity with tourists, while modern industry parks have proven to be the least sought-after category (SITPC, 2007c). Some other interesting general data: 11.83 per cent of the sites only receives groups, while the rest receives groups and individual visitors. Concerning entrance fees, 55.91 per cent of Shanghai's tourism attractions provides their services free of charge. Of all industrial tourism sites, 45.16 per cent is open to visitors all days of the year (whereby the biggest

Table 6.4 The six categories of Shanghai's industrial attractions

Category	#	Examples
Individual factories	18	Baosteel, Shanghai VW, SGM, Yakult
Modern industry parks	11	Free-trade zone, R&D park, Zhangjiang High-tech park
Creative industry clusters/parks	9	The Bridge 8, Tianzi Lane, M50
Industry museums and exhibition centres	33	Tobacco museum, ship-building museum
Industrial urban landmarks	10	Maglev, East-sea Bridge, Jinmao Tower
Others	12	Sony gallery, Oriental Land

Source: Annual Industrial Tourism Ticket Book (SITPC, 2007)

share of these sites are industrial museums); 39.78 per cent is open from Monday to Friday (mainly factories), 8.6 per cent is open from Monday to Saturday, and 6.45 per cent of the sites is only open on several special days. Normally, a tour takes less than 1 hour or 1–2 hours with the proportions of 39.78 per cent and 18.28 per cent respectively. 67.74 per cent of the sites has no booking requirements, while another 32.36 per cent only receives people who book the tour at least one week in advance (SITPC, 2007b).

4 Shanghai's industrial tourism products

In this section we analyse four selected industrial tourism attractions in Shanghai; our sample includes three individual factories (Baosteel, Volkswagen Shanghai and Yakult) and one creative industry cluster (M50). The attractions have in common that they all have been recognised as National Models of Excellence. They differ, however, in terms of ownership: Baosteel is a public company, Shanghai Volkswagen is a joint venture between the German Volkswagen Group and the public Shanghai Automotive Industry Corporation (SAIC), Yakult is a foreign private company and M50 is a company park in the legal form of a private company with Chinese ownership. Our selection covers the four legal forms of companies in China and two of the six categories of industrial tourism attractions as identified by SITPC. The three individual factories are the most visited industrial tourism attractions, together with the deep sea port of Shanghai, which counted 300,000 visitors in the first seven months of 2007[2].

Baosteel Group

The Baosteel Group Corporation is Shanghai's biggest and oldest state-owned company. Its headquarters and many important production facilities are located in the northern part of the province of Shanghai. The group is very famous throughout China and is widely associated with the city of Shanghai. In the 2004 Fortune Global 500 list, the Baosteel group occupied the 309th rank and thus is not only a Chinese economic heavyweight, but is also relevant on a worldwide scale. The steel manufacturer employs around 600,000 people in China and 14,000 people at its Shanghai location. The main products are different kinds of steel: intermediary products that are later on further processed by other companies (Baosteel Group, 2008).

The organisational foundations for Baosteel's industrial tourism programme were already laid in 1994, when the company decided to set up the Baosteel Travel Agency. At that time the agency was only intended for the arrangement of

2 The deep sea port tour has not been analysed in detail because it is not a National Model of Excellence and because only the port itself can be visited, not any other companies in the port area.

professional trips of Baosteel employees. In 1997 the agency proposed to launch company visits, hereby targeting tourists and citizens. This proposal was accepted by the Executive Board of the Baosteel Group. From its very beginnings the company's efforts to open up have been supported by the government of Shanghai, by local travel agencies and by the very positive feedback from local media. In 1999, the company was awarded with the 'Yearly Excellent Tourist Product Award' by the Shanghai Municipal Tourism Administrative Commission, and in 2004 it was the first company in Shanghai to be recognised as National Model of Excellence in Industrial Tourism.

Baosteel's factory tours take place from Monday to Friday from 9:00 a.m. to 16:00 p.m. and only a certain number of tours are allowed per day, in order to avoid disturbances of the daily operations. The holding capacity of the Shanghai plant has been limited to 100,000 visitors per year. Because of construction works, less visitors were admitted in 2007: in that year Baosteel received around 80,000 visitors including 6,000 foreigners. During certain sensitive periods (introduction of new products or new machines) parts of the factory or sometimes the full factory are off limits for company tours for several weeks, in order to avoid that visitors get in touch with company secrets. The company exclusively receives visiting groups from its own travel agency as well as from a couple of other travel agencies that have signed a cooperation agreement with SITPC. Sometimes individual visitors are joined with other groups.

Guided tours at Baosteel take around 1.5 hours and are also available in English (since 2008). A typical tour consists of the following elements: firstly, visitors get a guided tour through the Baosteel plant, then they visit the Baosteel Zoo (since 1987 a zoo exists within the Baosteel premises) and sometimes also a special outdoor entertainment programme is organised for them. If time permits, Baosteel then invites visitors to have lunch in the canteen, together with workers of Baosteel; this is done in order to make the visitor experience as authentic as possible. After lunch, visitors can participate in a guided visit of the Baosteel exhibition centre, where they get information about the history and development of Baosteel. This exhibition centre also aims to clarify the link between Baosteels' intermediary products and consumer products of everyday life. Baosteel's souvenir shop is the last stop for the visitors; this shop, with an annual turnover of almost €30,000, sells a wide array of souvenirs that are made of steel. In 2008 the company started an effort to improve and extend its tourist facilities: new facilities such as safe visitor corridors, a gathering centre, a restaurant, a telephone booth and a tourist toilet will be constructed. Also the employees of Baosteel benefit from these improvements. By the end of 2008 another attraction will be part of a visit of Baosteel: then, the Shanghai Steel Museum will open its doors to the public. This museum is financed by the Baosteel group and is also located on the Baosteel premises.

The Baosteel Travel Agency, which is owned by the Baosteel group and employs around 60 people, is the organisational unit that takes care of the organisation, planning and promotion of factory tours. This agency is also responsible for the training of the guides and for the concept of the tours. But the agency does not only organise the visits to the plant, it also promotes package tours. These package tours typically combine a Baosteel visit with leisure, shopping, sight seeing or with agricultural tourist attractions in the surroundings of the plant. Besides the travel agency, the Group also owns and operates three hotels (a 4-star, a 3-star and a 2-star hotel), all of which are located on the Baosteel premises. This is, however, nothing extraordinary in China, as many public companies operate their own guesthouses. In the case of Baosteel it was the peripheral location of the company that forced them to start operating their own hotels in 1980. Particularly during the summer months it is very difficult to find accommodation in the area for Baosteel's guests (including those who participate in a company tour). The 4-star hotel is used as a meeting and accommodation venue for professional visitors, and also accommodates the headquarters of the Baosteel Group.

The groups' industrial tourism visits are used as a showcase for the company's efforts in the fields of environmental protection and corporate social responsibility. On Baosteel's premises, a greenbelt covers 42.71 per cent of the whole area (Baosteel Group, 2008). The Baosteel Zoo not only attracts tourist, but is also used as a showcase to convince the Baosteel staff and citizens in the immediate neighbourhood that the area in and around Baosteel is a safe and healthy place to be. Another reason to organise company tours is to build loyalty among citizens and tourists by expressing the company's close connections with their daily lives. This is achieved in a rather indirect way: in its exhibition centre, Baosteel explains which consumer products contain their steel products. For Baosteel it is important that people who (could) work as purchasers at a corporate customer hold a positive picture about the company, as positive effects of this on business are expected. Baosteel invites the employees of their corporate customers to participate in a tour, and also some of the students that visit Baosteel might in the future be employed with companies that are potential or current customers of Baosteel. Furthermore, Baosteel sees industrial tourism also as a good way to increase the loyalty of their shareholders. Since 2002 Baosteel issues its shares also on the stock market. Baosteel is convinced that the constant presence of visitors encourages workers to work harder; thus they use this as a way to control their own employees. The fact that Baosteel has a kind of exemplary function for other companies, is also seen in a positive way: many other companies visit Baosteel in order to learn how to implement industrial tourism in their own firms. The Baosteel group considers this an outstanding opportunity for network building and to get in contact with potential customers.

Shanghai Volkswagen

Established in March 1985, Shanghai Volkswagen (hereafter, SVW) is the first joint venture of a foreign car manufacturer in China. Located in the north-western outskirts of Shanghai, SVW operates three car plants and an engine factory on the site. SVW produces mainly passenger cars and has an annual production capacity of 500,000 units (Shanghai Volkswagen, 2008). The factory produces models such as the Santana and the fifth generation of the Passat. Due to lower labour costs the factory is less automated than similar plants in Europe.

Already in 1985, when the plant was set up, SVW built a visitor centre for receiving professional tourists. In this initial phase the influx of visitors was rather limited, which was also due to a lack of workforce in the department to organise the visits. It was only in September 2003 that Shanghai Volkswagen started to offer industrial tourism visits also for the wider public. Since then, SVW is open for tourists and citizens from 9:00 a.m. to 16:00 p.m. from Monday to Friday. To organise the visits SVW exclusively cooperates with a private travel agency (Shunlian Travel) which employs 30 people. The agency is fully responsible for the organisation of tours and reception of leisure visitors, as well as for the promotion of these visits. The agency is also responsible for arranging corporate trips for Volkswagen employees, as well as for the arrangement of internal business meetings and conferences. The exclusive contract also implies that the travel agency is not allowed to arrange trips for any other customers or to run any side businesses of whatever kind. Volkswagen has retained the responsibility for professional visitors within its own organisational structure. This so-called Affairs Management Office is responsible for the organisation of professional visitors and VIPs. Apart from organising tours, this department is also responsible for the communication with VIPs, government officials and politicians in general, as well as for the provision of conference facilities for business meetings.

Today the annual holding capacity for visitors has been set at 100,000 per year. This capacity has been fully reached over the past years. Due to the huge demand, interested parties have to arrange their visits several weeks ahead. SVW provides a standard tour of 1 hour duration for its visitors, which is also available in English. Visitors of SVW usually start their visit with an introduction movie about the history and tasks of the company. After the movie, a kind of pick-up train takes visitors through one of the three car plants (it is not possible to visit the engine plant). Similar to Baosteel, also SVW puts certain areas temporarily off limits for visitors, if a new production line or new machines are introduced. During the tour, headphones are provided to the visitors, in order to be able to follow the explanations of the guide in the sometimes noisy factory. The drivers of the trains also act as tour guides (while they are driving). This, coupled with the fact that the visitor train rarely stops and a big amount of information is conveyed within the given timeframe, limits the possibility to interact and to ask questions. The visit ends with a visit to the SVW exhibition hall, where the finished car models produced in Shanghai are shown, and where visitors can buy souvenirs and take pictures (it is forbidden to take photos inside the plant). In its communication to visitors, SVW emphasises its concern for their employees and for the environment, as well as for the quality of the company's products. The slogans that concern the issues mentioned have been placed on visible locations along the route of SVW plant visitors.

The reasons for SVW to launch industrial tourism are manifold, whereby strengthening the company image and brand and increasing the number of cars sold are the most important ones. As SVW is a public-private joint venture, also the public side has its share of influence on the industrial tourism strategy. Shanghai Volkswagen is also intended to be a showcase to demonstrate to Chinese citizens and the outside world alike what kind of achievements China has obtained over the last decades.

The visit programme of Shanghai Volkswagen is oriented towards the plant visits of Volkswagen's parent company in Wolfsburg (Germany), which is clearly seen as an inspiring example for SVW. Delegations of SVW have visited the Volkswagen plant in Wolfsburg several times and SVW also recognises that still a lot can be learnt from the Wolfsburg factory in many respects. SVW indicated that for the future an extension of the existing visitor facilities is planned. Moreover, SVW has plans to promote itself as one element of an overall tourism package, together with other attractions in its surroundings. Nearby attractions that come into question are the old town of Anting (20 minutes by car), the Formula 1 circuit (15 minutes by car), the car museum (5 minutes by car) or the theme park like 'Auto City' (a 10-minute drive).

M50 Creative Park

The M50 Creative Park is located on the banks of the Suzhou Creek in the city centre of Shanghai and accommodates fashion, cultural and other creative companies (artist studios, art galleries, film producers, educational institutes in the field of art, fashion designers, architects, urban planners, product designers and so on). The M50 premises had formerly been occupied by the Shanghai Chunming Roving Factory, a well known textile company which is generally considered to be the cradle of modern manufacturing in Shanghai. The ensemble of buildings on the M50 premises is an industrial heritage site: the site was built up from the 1930s until the 1990s and thus features historic buildings from various periods. In the late 1990s, the textile factory had to close its operations and most of its employees had to leave. The company tried to retain some of its employees and therefore decided to set up the M50 Real Estate Management Company, where 20 persons from the original factory are still employed today. This company redeveloped the former factories and warehouses on the site, with the aim to rent the space to companies. The M50 company park opened in April 2005 and by the end of 2007, more than 100 companies from 16 different countries had moved into the office space provided by M50 (M50 Creative Park, 2008).

M50 is quite a special location for industrial tourism, as it combines industrial heritage with company visits on one site. The origins of the park's industrial tourism initiative are to be found in the year 2006, when a delegation from Shanghai's Municipal Tourism Administrative Commission and of SITPC visited the park. After several meetings with these actors, M50 decided to open its premises for industrial tourists. Since 2007, M50 organises company visits to the businesses on the site. The Planning and Administration Department of the M50 Real Estate Management Company is in charge of designing, planning, managing and promoting the visits to the company park. An outcome of this cooperation is the introduction of an 'obligatory open door policy' for companies in the park: companies are only admitted to the M50 park if they agree to open their doors for group visitors. According to this contract companies have the right to reject single visitors, and of course, the companies can decide to keep certain areas off-limits or close on certain days if the visitors would bother their daily operations or if visits could expose important company secrets. M50 periodically organises free training classes in order to transfer the needed know-how for company visits to the companies in the park. Companies that intend to move into the M50 park are expected to participate in such a training, even before they start their operations. This also ensures that the companies take the necessities of visitors into account when designing the interior of their workshop or office. To give an example, many companies have decided to open a kind of exhibition space for visitors in the wake of these trainings.

The M50 company park is open to the public from Tuesday to Sunday, from 10:00 a.m. to 17:00 p.m. By the end of 2007, M50 could mainly attract professional visitors such as art dealers, artists and art students. M50 then started to put more emphasis on attracting citizens and especially people from their immediate neighbourhood. In 2008, M50 managed to draw an average of 400 visitors per day, of whom, admittedly, only a minor part participated in a guided tour (China Art Networks, 2008). That is to say, visitors can enter the M50 area free of charge and if a company has an exhibition space, this area can be visited without a guided tour during opening times. On the other hand, the workshops and offices are only accessible in the course of a guided tour. Guided tours have to be booked on prior notice (at least one week ahead). Furthermore, M50 operates its own general exhibition hall where souvenirs and products of companies that are located in the park are sold.

M50 had important reasons to become active in industrial tourism: the company visits are intended to promote the M50 brand, and in this way also to increase the visibility of the single company brands, which is expected to lead to economic benefits. Especially for the artists in the park benefits are expected, even in the short run, simply because they might be able to sell more works of art. For the other companies benefits are expected in the medium and long run. M50 also recognises that industrial tourism constitutes an important opportunity for the member companies to engage in networking. For the future, the district authority of the Suzhou Creek plans to develop a visitor package that will bring together attractions in the area, among which M50. M50 supports these ideas. Potential elements of such a package are rather conventional tourist attractions, but as well nearby sites that do have a relation to industry, such as the Shanghai Cotton and Textile Museum and the Shanghai Beer Museum.

Yakult

Yakult Honsha CO.LTD. was set up in 1955 in Japan and is nowadays one of the biggest producers of probiotic drinks (lactic acid bacteria drinks) worldwide. Yakult operates 21 factories in Japan and 14 factories abroad. In August 2006 Yakult opened a plant in Shanghai's Jiading District, in the north-western part of the city. This plant produces 300,000 bottles of probiotic drinks daily, and supplies an area of 46,000 km^2 with its products (Yakult, 2008).

The Japanese parent company has a long tradition in industrial tourism: the necessities for visitors had already been taken into account during the planning phase of the Shanghai plant, for example by including visitor toilets, a reception room and a lecturing room. The most outstanding feature of all Yakult factories worldwide is an elevated corridor for visitors. This is a kind of closed glass walkway that runs through the whole factory and allows visitors to get a glimpse on all steps of the production process, also on steps of production that would have been off limits normally, for the sake of hygiene. Already in the planning phase of the Shanghai factory (at the end of 2004), Yakult approached the Shanghai municipality and expressed its interest to be involved actively in Shanghai's industrial tourism initiative. Yakult started to organise company visits right after the opening of the plant in August 2006. The plant is open for leisure visitors from Monday to Friday, from 9:00 a.m. to 16:00 p.m. These tours are organised for free and Yakult even bears the costs for shuttle buses from the city centre of Shanghai. From August 2006 to October 2007, more than 80,000 visitors had participated in a company tour at Yakult Shanghai.

Yakult regards kids, families, retired people, local citizens, business professionals and government officials as their main target groups. The firm approaches these groups and invites them to participate in a company tour. Accordingly, the company offers three standard visitor programmes: one for children and families, one for adults (retired people and local citizens) and one for professionals and government officials. Generally speaking, the visit programme includes an introduction movie about Yakult (available in Chinese, English and Japanese), followed by a presentation about health and the role of Yakult in health care. While for the adult group this presentation includes a lecture of how to lead a healthy life, children receive a somewhat more interactive programme. With the help of two employees that are dressed up like comic figures, Yakult provides an entertaining, yet educating programme in which the good (Yakult) bacteria fight against the bad bacteria.

After this introduction, Yakult hands out a bottle of its probiotic drink to every visitor. Then, the visitor groups enter the glass corridor where they can witness the production process of Yakult. The glass corridor can accommodate up to 200 visitors at a time. In the past Yakult only provided one guide for 200 visitors in some cases, which led to complaints by visitors. At this time Yakult employs four to five guides for a group of this size. If there are a lot of visitors at the same time, Yakult also allows visitors to have a look at the water treatment plant and the cold-storage room that belongs to the factory. This is mainly done to distribute visitors more evenly in space and to avoid congestion. At the end of the visit, there is a possibility to buy Yakult's probiotic drinks at 10 per cent discount on the market price. Altogether, the whole programme takes around one hour. Yakult plans to include also its logistic centre, into the tour, as well as two facilities that have to be constructed yet: the Yakult museum and an exhibition centre about the company.

The Public Relations office of Yakult Shanghai (a sub unit of the marketing department) is the leading force in the design and organisation of factory visits. People that are interested in a company tour can arrange their visit directly through this department. Visits can alternatively be arranged via one of the cooperating travel agencies in Shanghai. Yakult spends around €100,000 on the organisation of company visits, including the salaries of the guides and the shuttle transport from the city centre. This investment comes from the company's marketing budget. The importance of company visits as a promotion tool and a tool for image building has been increasing constantly over the past few years, which is also reflected by the distribution of Yakult's marketing budget: the budget for company tours has been increasing annually while the annual budget for traditional advertising has been decreasing accordingly. Yakult's preference for company tours as a marketing measure is rooted in the observation that a TV spot only gives you on average 15 seconds to convey your message, while a company visit provides you with one hour to reach your customers. Visitors often express their astonishment about the cleanness of the plant, and this is also one of the things that Yakult wants to demonstrate. Yakult is convinced that the company visits have led to an increase of sales, even in the short run.

5 Applying our frame of analysis

In the previous two sections we analysed the organisational framework of industrial tourism in Shanghai, followed by a more detailed description of four National Models of Excellence. In this section we apply our frame of analysis to the observations we made, trying to find out under what conditions industrial tourism can be beneficial for the strategies of companies and local authorities. We subsequently deal with 1) adequate target group selection; 2) the attractiveness of the product; 3) benefits in terms of income and image for city and enterprise; and 4) the organisational setup.

Adequate target group selection

An important aspect of industrial tourism development is adequate target group selection. With six key target groups and six types of industrial tourism attractions the government of Shanghai has identified a relatively large market for industrial tourism, explaining the impressive visitor figures. Interestingly, leisure tourists are not among the main target groups for industrial tourism in Shanghai, and also their share in the overall number of industrial tourists is very small. This can be explained by the fact that leisure tourism is not rooted in the Chinese (travel) culture. There are fundamental differences between China and Europe in terms of tourist behaviour: the Chinese outgoing tourist market is still in a very early stage and Chinese tourists are rather inexperienced. Unlike Europe, where tourists act more and more individually and tailor made trips are becoming more popular, this trend is still in a very embryonic phase in China. If Chinese people travel, they usually do so for the purpose of visiting relatives, shopping or for business reasons. In China, individual leisure tourism as such hardly exists and most people still prefer to travel in groups.

Most individual companies do not target all groups identified by the local government. Each company has specific interest in industrial tourism and accordingly addresses specific target groups. It is not common, however, to exclude groups from a visit: most firms emphasise that they welcome all different kinds of visitors, no matter their background or age. Some companies do exclude individuals, but that is mainly because of financial-organisational reasons (it is more expensive and not efficient to receive individuals). In our sample of four companies we observe that firms pay special attention to business visitors, by arranging tailor-made tours. This is most obvious at SVW where the organisation of tours for visitors with a leisure and educational motive has been outsourced to a travel agency, while visitors with a political and business motive, who are basically more important for the company's core business, are received by the company itself. Only the M50 company park, which boasts a number of companies that produce consumer products, has recognised single leisure visitors as a very important target group.

Also with regard to the scale of the target groups, differences can be observed between the city (SITPC) and the individual companies. SITPC claims to target an international audience, although its website is in Chinese only. Some individual companies are clearly a step ahead: an increasing number of firms is providing tours in English (like Baosteel did in 2008), and is developing special VIP tours for foreign guests with a business motive (SVW being one example). Apart from international guests, also local visitors get more than average attention from companies, but that is in line with SITPC's strategy.

One of the target groups that has been identified by SITPC are families. In practice, however, it appears to be difficult to attract the middle-aged population, as statistics demonstrate that young people (10–25 years) and older people (50+) are overrepresented. The age group in between (25–50) is particularly interesting for both city and enterprise as people who are economically active can be expected to have a bigger multiplier effect (a bigger network), thus making a larger contribution to changing the image of Shanghai and the companies. Also from an income point of view, the middle age group probably generates more benefits than young people visiting the firms: they spend more money in hotels, restaurants and shops. One of the instruments SITPC has developed to attract this highly interesting group ('closing the age gap') is the Annual Ticket Book: many employers now give away the Annual Ticket Book to their employees as a sort of incentive. It is not yet clear, however, if this instrument has already changed the age structure of the industrial tourist population.

The attractiveness of the industrial tourism product

The second issue we want to address is the attractiveness of the industrial tourism product. SITPC has carried out a market survey to find out what kinds of products industrial tourists appreciate most: it appeared that they prefer visiting creative parks and big factories. Furthermore, it became clear that companies that do not produce consumer goods, like Baosteel, can be very appealing to visitors, as long as these companies are somehow related to the identity of a city. Several companies in Shanghai have made investments to increase the attractiveness of their visit, notably by improving visitor facilities such as a reception, toilets, lecture rooms, tourist corridors, and industrial museums. In general, tours have an educational character, leaving relatively little room for interaction. This is understandable in view of the Chinese culture, but may be perceived as less attractive to foreign visitors. An important ingredient of the attractiveness of industrial tourism products is authenticity: Baosteel secures this aspect by inviting visitors to have a lunch together with the Baosteel workers in the company cafeteria. Furthermore we observed that the four companies we analysed in detail have increased the attractiveness of a visit to their site by linking up with surrounding leisure, shopping and traditional sightseeing attractions.

SITPC plays a key role in securing and improving the attractiveness of Shanghai's industrial tourism product. The centre provides advice and developing quality standards, which also stimulates companies to renew and innovate their programme, and also issues the Annual Ticket Book, which has increased the visibility and accessibility of industrial tourism attractions. Moreover, the website and other promotional tools have contributed to the increase in visitor numbers. The Annual Ticket Book stimulates holders to visit more companies in one year, possibly also by combining two visits on one day. The Annual Ticket Book also fits in a strategy to create critical mass: by integrating all attractions into one product, it is easier to reach a larger audience, which in fact consists of different target groups with different preferences.

In terms of physical accessibility Shanghai finds itself in a challenging situation: most manufacturing companies are located in the suburbs of the city and, due to a lack of public transport, are rather detached from the city centre. New metro lines have been and will be developed, but the metro system is not yet as large as in cities like London or New York. This creates a big obstacle for leisure tourists who do not have their own mode of transport. Some companies have recognised this problem and started to operate connections with their own shuttle buses for tourists (for example Yakult). Other companies rely on shuttle buses that are offered by travel agencies. The importance of distance as a barrier is also reflected in the zoning plan for industrial tourism: by developing critical mass of industrial tourism attractions in the seven designated areas, and by strengthening the link of these attractions with conventional attractions, these areas are to become more lively and are to generate conditions under which public transport connections are economically feasible.

Benefits in terms of income and image for city and enterprise

The third condition for industrial tourism development is the occurrence of tangible and intangible benefits for city and enterprise. As far as tangible benefits for the participating firms are concerned, none of the four companies we analysed receives any income through entrance fees. Three firms, Baosteel, M50 and SVW, indicated that they do earn some money mainly by selling souvenirs and products (the latter only in the case of M50 and SVW). But these revenues are very small and cannot compensate for the costs incurred for arranging the company tours. Since all four companies spend considerable – and in some cases (like Yakult) increasing – amounts of money on industrial tourism, it is highly probable that the return on investment is mainly generated by intangible benefits. From the interviews we conclude that managers of these companies are convinced that industrial tourism programmes can bring substantial indirect benefits, ranging from increases in sales, to strengthening the image of the company, increasing customer loyalty and making it easier to attract qualified employees. Each company, however, places the emphasis on a different (set of) intangible benefits: Yakult regards industrial tourism as a good way to promote the products in the short run and to strengthen the company brand in the long run; this method is even considered to be more effective than traditional advertising methods. Both Yakult and M50 expect a short term increase of sales as a result of company tours, while steel manufacturer Baosteel uses industrial tourism to show citizens and employees that the company does everything to limit health- and environmental risks, and to increase the loyalty of business customers.

Shifting our attention to the benefits for the city, we can state that industrial tourism makes a small but substantial contribution to the total annual revenues of tourism in Shanghai. In 2006, just after the foundation of SITPC, the share of industrial tourism was already 6.4 per cent (SITPC, 2006). The city of Shanghai has recognised industrial tourism as an adequate tool to satisfy visitors that have already been to the city once or more times. Many of these returning visitors come for business meetings, and Shanghai considers it especially important to please this group (this is why information material about industrial tourism is mainly displayed in upmarket hotels). This makes sense, as many individuals of this group hold influential positions and can be considered to be opinion leaders. If some of these key persons can be convinced that there are more reasons to visit Shanghai apart from shopping and business, this could be an important step for Shanghai to be perceived as city that is able to satisfy a diversified range of touristic demands. In this respect, industrial tourism is both a way to diversify the range of attractions in the city and a catalyst to diversify the one sided image visitors hold about Shanghai. Moreover, one sub-group of the foreign business visitors can be especially fruitful for a city: investors. Industrial tourism can be used as a tool to attract inward investment. If industrial tourism is organised in a professional way it may help to convince potential investors about the attractiveness of the city and/or district, and thus can contribute to the development of further economic activities in the city.

In China the interests of city and enterprise are often converging as many companies are (partly) owned by the government. This implies that it is difficult to separate benefits for companies from benefits for Shanghai and China. In the development of industrial tourism, the state has played a key role with its National Models of Excellence programme which clearly aims to further improve the image of China as a modern nation. Companies that are owned (and in some kind of way also managed) by the government may invest in industrial tourism in view of objectives that private-owned companies would never think about. The most obvious example in this respect is Baosteel which has played a key role the last decades in representing the new China, but also joint ventures like SVW are more or less expected to participate in China's marketing campaigns.

The organisational setup

The fourth condition is the organisational setup, which relates to the way industrial tourism is organised within the four companies, as well as the organisational framework developed on the level of Shanghai (SITPC).

With regard to the internal organisational structures we identified various models that all seem to work well, which leads us to the conclusion that there is not one organisational solution that fits all companies. Baosteel and SVW both make use of a travel agency, which is charged with the arrangement of industrial tourism visits. While Baosteel's travel agency is an organisational unit within the Baosteel corporation, SVW has out-contracted the organisation of company visits to an external agency (and has only retained the responsibility for professional and VIP visitors within SVW). This travel agency is exclusively responsible for industrial tourism and for the arrangement of corporate trips. In contrast to that, the agency that is owned by Baosteel operates like a regular travel agency that also organises trips and excursions that have no link at all with industrial tourism or with the corporate travel necessities of Baosteel employees. In the case of Yakult, the public relations section (part of the marketing department) is responsible for the organisation of visits. This can be explained by the fact that Yakult regards industrial tourism purely as a marketing measure; and an important one, actually. M50 is a special case when it comes to the organisation of company visits: as owner of a company park, this organisation is responsible for the organisation of guided tours to all the companies that are located in the park.

Taking a look on the organisational environment of industrial tourism in Shanghai, we can state that the local and national governments managed to create an environment that is conducive to industrial tourism. The local government planned and financed the foundation of an organisation (Shanghai Industrial Tourism Promotion Centre, SITPC) that embraces a wide range of tasks and stakeholders in the realm of industrial tourism. Together with a national policy ('national models of excellence') this has spurred the development of industrial tourism in Shanghai. The SITPC has been created in a top-down approach and is designed as an all-embracing organisation for industrial tourism. Due to the different cultural and political context in Europe, it might be very difficult to realise the same type of organisation there. Nevertheless, certain aspects of the SITPC are also highly interesting for the European context: for example the quality control function, the consultancy function and the training function. Also the data sharing system that links all participating companies and the annual industrial tourism ticket can be considered best practices.

References

Baosteel Group 2008. [Online]. Available at: http://www.baosteel.com [accessed: 2008].

China National Tourism Administration 2008. [Online]. Available at: http://www.cnta.gov.cn/lyen/ [accessed: 2008].

China National Tourism Statistics 2008. [Online]. Available at: http://www.cnto.org/ [accessed: 2008].

China Art Networks 2008. Shanghai SoHo. [Online]. Available at: http://www.chinaartnetworks.com/feature/wen21.shtml [accessed: 2008].

M50 Creative Park 2008. [Online]. Available at: http://www.m50.com.cn [accessed: 2008].

Shanghai Institute of Tourism 2007. *Shanghai Industrial Tourism Development Plan 2006–2010*. Shanghai: Shanghai Institute of Tourism.

Shanghai Municipal Government 2008. [Online]. Available at: http://www.shanghai.gov.cn [accessed: 2008].

Shanghai Statistics Bureau 2008. [Online]. Available at: http://www.stats-sh.gov.cn [accessed: 2008].

Shanghai Volkswagen 2008. [Online]. Available at: http://www.csvw.com [accessed: 2008].

SITPC 2007a. Shanghai Industrial Tourism Promotion Centre, www.itripsh.gov.cn [accessed: 2008].

SITPC 2007b. Annual industrial tourism ticket book.

SITPC 2007c. Market survey on industrial tourism.

Statistical Yearbook 2007. Shanghai Statistical Bureau, [Online]. Available at: http://www.stats-sh.gov.cn/2004shtj/tjnj/tjnj2007e.htm [accessed: 2007].

Yakult 2008. [Online]. Available at: http://www.yakult.com.cn/factory/ [accessed: 2008].

Discussion partners

Ms. Li Fengfeng, Shanghai Industry Tourism Promotion Centre.
Prof. Dr. Gaojun, Vice President of Shanghai Institute of Tourism, Shanghai Normal University.
Mr. Wang Jianming, Director of Policy & Law Department, Shanghai Municipal Tourism Administrative Commission.
Mr. Shen Jianping, Manager of Comprehensive Department, Shanghai International Traveling Agency, Baosteel Group.
Ms. Candy Li, Supervisor of Public Relations Department, Bridge 8.
Mr. Lin Zhan Lin, Policy & Law Department, Shanghai Municipal Tourism Administrative Commission.
Mr. Tang Linlin, Secretary, Shanghai Industry Tourism Promotion Centre.
Ms. Liu Qing, Planning Department, Shanghai Industry Tourism Promotion Centre.
Mr. Hua Yao Sheng, Manager of Domestic Tourism Centre, Shanghai International Travel Agency, Baosteel Group.
Mr. Jin Weidong, Chief Manager, M50.
Mr. Wang Yijun, Director of Public Relations Department, Yakult Shanghai.
Mr. Guo Zhongnan, Director of General Manager Office, Shanghai Volkswagen.

Chapter 7
Industrial Tourism in Rotterdam

1 Introduction

Rotterdam was one of the first cities on the European continent that discovered industrial tourism as a niche market in urban tourism: already in 1997 the local tourism organisation started to offer visits to operational firms. In 2003, however, political and financial support for the development of industrial tourism declined, and subsidies to industrial tourism development were cancelled. Today several organisations are somehow involved in supplying company visits, but the general impression is that Rotterdam is not fully utilising its potential, with one of the largest ports of the world as a unique asset.

In this chapter we analyse the industrial tourism supply of Rotterdam (section 4), and the organisational framework for industrial tourism (section 3). Before we introduce the organisations and companies that are active, we first give a general description of the city (section 2), in which we pay particular attention to the city's economic structure and tourism profile. After that we apply our frame of analysis to the observations we made in Rotterdam (section 5), resulting in some conclusions.

2 General profile of Rotterdam

History and demographic development

Rotterdam was founded as a small settlement along both sides of the Rotte river, currently known as a tributary of the Nieuwe Maas. After it was awarded city rights in 1340, the town first developed into a medium-sized trading city. From the 17th century onwards, the city gained further importance, starting to benefit from its location near the sea. The population grew from 30,000 in 1650 to 300,000 in 1900. In the second half of the 19th century, Rotterdam started to expand its port, with the development of several port areas (Maashaven, Rijnhaven and Waalhaven). In the Second World War the city was bombed resulting in the devastation of nearly all historical buildings in the city centre (on the north bank of the Nieuwe Maas).

After the war the city was rebuilt in what could be called a North American style, starting to construct the tall buildings that dominate the city's skyline today. Since the 1960s port activities gradually moved toward the sea, as in many other port cities, with the development of Europoort, the Botlek Area and the Maasvlakte. As a consequence former port areas in the city have become available

Table 7.1 Employment structure in Rotterdam, compared with other Dutch cities (2006)

Sector	Rotterdam	Amsterdam	The Hague	Utrecht	Average
Agriculture	0.17%	0.21%	0.55%	0.10%	0.26%
Industry	8.38%	4.48%	3.96%	4.63%	5.36%
Utilities	0.73%	0.13%	0.13%	0.00%	0.25%
Construction	5.73%	2.96%	2.60%	4.22%	3.88%
Trade	11.95%	13.28%	10.25%	13.56%	12.26%
Hotel and catering	3.00%	5.34%	3.16%	3.23%	3.69%
Transport/ communication	11.66%	6.81%	8.87%	6.11%	8.36%
Financial services	5.32%	9.84%	4.76%	6.83%	6.69%
Business services	18.05%	23.75%	15.93%	24.14%	20.47%
Government	6.03%	7.06%	23.32%	6.06%	10.62%
Education	8.10%	6.49%	5.31%	10.29%	7.55%
Health and welfare	16.38%	12.98%	14.47%	15.73%	14.89%
Culture and other services	4.48%	6.67%	6.70%	5.08%	5.73%

Source: Kerncijfers Rotterdam (COS, 2007), Bedrijvenregister Zuid-Holland, O+S, Utrecht and the Hague

for redevelopment. Since the 1990s the city is developing the Kop van Zuid, a city-centre expansion on the southern bank of the Nieuwe Maas river, with the Erasmus Bridge as characteristic landmark. After 2010, the port will be further expanded with Maasvlakte 2, by regaining land from the sea. The first part will be ready for use in 2013. The distance from the city to this new area is about 50 km.

The population of the city reached its peak in 1965 with 732,000 residents. Since then, suburbanisation resulted in a decline to less than 600,000 inhabitants. For some years now, the population number is relatively stable. An important characteristic of the Rotterdam population is the high share of ethnic minorities. Almost half of the population has a foreign background (COS, 2007), with approximately 170 nationalities being represented.

Location and accessibility

The city of Rotterdam is located in the so-called south wing of the Randstad, a conurbation of the four largest cities in the Netherlands – the others being Amsterdam, The Hague and Utrecht – and several smaller ones. Although the Randstad has no official status, it is often considered a spatial-economic entity comparable with the German Ruhr Rhine region. The south wing of this ring-shaped metropolitan region includes the regions of Rotterdam (1.2 million inhabitants) and The Hague (1 million inhabitants), as well as the cities of Leiden and Dordrecht (both with more than 100,000 inhabitants). The straight-line distance between the centres of Rotterdam and The Hague is less than 25 km. The distance between Rotterdam and the other two major cities – Amsterdam and Utrecht – is considerably larger, for both more than 60 km.

The international accessibility of Rotterdam is very good. The city has its own regional airport, with more than 1.1 million passengers in 2007 (Rotterdam Airport, 2008) travelling from and to more than 25 destinations in Europe. Moreover, Rotterdam is one of the stops of the high-speed train from Amsterdam to Brussels and Paris (and London via Brussels). With the new rail track between Amsterdam and Rotterdam, the connection with Schiphol Airport – the fifth largest airport of Europe – and Amsterdam's city centre has been further improved, with travel times of half an hour and 40 minutes respectively. Rotterdam is well positioned in the Dutch highway system with direct connections to The Hague and Amsterdam (A13, A4), Utrecht (A20, A12) and several cities to the south and southeast (A16).

Internal accessibility is secured through the ring road (A20, A16, A15) around the city. During peak hours, however, congestion is a serious problem (as elsewhere in the Randstad). Between 2009 and 2015 the capacity of the southern part of the ring road (A15) will be improved, which will enhance accessibility on the long run, but might reduce accessibility on the short run (because of construction work going on). In the city's infrastructure the Nieuwe Maas river has always acted as a barrier between the northern part and the southern part of Rotterdam. The city centre, which includes the area around the central station, is located to the north of the river. Several tunnels and bridges (Benelux Tunnel, Maas Tunnel, Erasmus Bridge, Willems Bridge, Willems Tunnel, Van Brienenoord Bridge) have been constructed to reduce the river's barrier effect.

Public transport provides an alternative for travellers: the city has a tram and metro system which is connected to Randstad Rail: a lightrail system connecting Rotterdam and The Hague. In addition, there is a water bus system which provides regular services, notably for commuters, between Rotterdam and Dordrecht (located downstream). Recently, the city introduced a fully-fledged public transport system over water with several landing stages along the river, from Hoek van Holland (upstream) to Krimpen aan den IJssel (downstream).

Economic structure

Rotterdam's economy has for a long time been dominated by the port and industrial complex. With a goods throughput of more than 400 millions of tons, the Port of Rotterdam is still the largest port of Europe (Annual Report, Port of Rotterdam, 2007). It used to be the largest port of the world for several decades (in terms of throughput), until Shanghai and Singapore took over this leading position in 2004.

In 2000, it was calculated that the port provides direct employment to 60,000 people (in 2006 the port generated 86,500 jobs according to the 2007 annual report), and indirect jobs to an additional 250,000 people, however not necessarily to Rotterdam citizens (this applies to both numbers) (Port of Rotterdam, 2008). Looking at the employment structure of Rotterdam, it appears that the share of industrial jobs is considerably higher than in the three other big cities. Nevertheless, also in Rotterdam the percentage of industrial jobs has considerably decreased in the last decades, with more jobs emerging in the service sector. Among these new jobs are also jobs that are somehow related to the port and industrial complex, for instance when it comes to providing business and financial services to logistical firms. The percentages in the table above also make clear that Rotterdam is relatively strong in transport and communication, construction, and health and welfare. Less important, in terms of employment, is the hotel and catering industry, indicating that Rotterdam is not a typical tourist destination (notably if compared with Amsterdam).

With almost 17,000 jobs the municipality (City of Rotterdam) is the largest employer of the city, followed by the Erasmus Medisch Centrum (hospital) with about 9,500 jobs (see the table below). The most important private-sector employer is Unilever, a British-Dutch producer of home, personal care and foods products with well-known brands such as Lipton, Knorr, Dove and Omo. One of the two corporate headquarters of this company listed on the Fortune 500 (122nd position[1]) is located in Rotterdam, in a eye-catching building ('the bridge') right at the river Nieuwe Maas, on top of the Blue Band (butter) factory. In the Rotterdam region, Unilever also had factories in Vlaardingen (cleaning products) and Delft (peanut butter and mayonnaise), but these factories were closed in 2008. The production of peanut butter (Calvé) was relocated to the factory in Rotterdam, while other activities moved away to other parts of Europe. What remained in Vlaardingen, however, is the Foods and Nutrition Research Institute with about 900 jobs. In 2007, Unilever provided employment to almost 3,500 people in the region (OBR, 2007). Another large company with its roots in the Rotterdam region is Royal Dutch Shell, one of the three largest companies in the world according to Fortune 500. Since the early 1990s, however, this British-Dutch company is headquartered in The Hague. Today the presence of Shell in Rotterdam is limited to the largest oil refinery of Europe and a chemical complex in Pernis. Shell Pernis provides employment to 2,250 people (Shell Pernis, 2008). The second-largest oil refinery of Europe is also located in Rotterdam and operated by BP, providing employment to 700 people. Another big company in the port is ECT B.V. (Europe Container Terminals) which operates three big container terminals in the port of Rotterdam and generates more than 2,000 jobs. Since 2004, ECT B.V. is owned by the Hutchison Port Holdings Group (HPH), with headquarters in Hong Kong. With 1,250 jobs the Port of Rotterdam (the authority) is also an important employer, worth mentioning (OBR, 2007, p. 30). Other industrial and port-related companies

1 http://money.cnn.com/magazines/fortune/global500/2008/full_list/.

that are headquartered in the Rotterdam region are Hunter Douglas (window coverings), LyondellBasell (chemical industry), Damen Shipyards (shipbuilding and repair), SBM Offshore in Schiedam (formerly known as IHC Calland, oil and gas industry), Boskalis Westminster in Papendrecht (dredging services), Vopak (storage of oil and other bulk liquids), IHC Merwede in Sliedrecht (dredging services and more), Smit Internationale (maritime services) Mammoet in Schiedam (heavy lifting and transport), Refresco in Dordrecht (formerly known as Menken, beverages), AVR (recycling), Meneba (grain), Keppel Verolme (offshore conversion and modification, ship repair) and Van Uden (stevedoring).

Table 7.2 Selection of the largest employers in the Rotterdam region (January 2007)[2]

Company/ organisation	Sector	Headquarters	Employees
City of Rotterdam	Administration	Rotterdam	16,940
Erasmus Medisch Centrum	Health	Rotterdam	9,430
Unilever	Industry (food, care)	Rotterdam	3,460
ING Nederland	Services	Amsterdam	3,370
Medisch Centrum Rijnmond Zuid	Health	Rotterdam	2,910
Eneco	Energy	Rotterdam	2,650
Ahold	Retail	Wormer	2,460
Imtech	Technical services	Gouda	2,080
ECT B.V.	Logistics (containers)	Rotterdam	2,070

Source: Economische Verkenningen Rotterdam 2007 (OBR, 2007), own selection

Unemployment is relatively high in Rotterdam, with about 9 per cent of the labour force having no job (2006 data, OBR, 2007), compared to 5 per cent for the Netherlands. At the same time, however, there are also many job openings, notably in the port and industrial complex. This situation is not only the result of the current economic tide, but also of a more structural discrepancy between the supply of, and demand for labour. Not only in the service sector, but also in the industrial sector, there is a strong demand for well-skilled and well-educated personnel. In comparison with Amsterdam, The Hague and Utrecht, the average educational level of the population and their degree of participation is low, notably

2 This is only a selection of the largest employers meant to give an impression of what companies are located in the Rotterdam region. Not included are, for instance, the police, the Rotterdam transport authority, and several healthcare institutions. Also Shell is not included here, although this company provides employment to 2,250 people.

among well represented ethnic minorities (e.g., Turkish and Maroccan). About 30 per cent is higher educated, a considerably lower share than in Amsterdam (almost 50 per cent), Utrecht (more than 50 per cent), and The Hague (almost 40 per cent).

Rotterdam is not a typical student town, like Amsterdam, Utrecht or Leiden, but has a rich offer of higher education, with the Erasmus University (Economics, Medicine, Business, Law, etc.) as one of the strong assets, and 55,000 students in total. One of the aims of the City (the municipality) is to create a more attractive living environment for higher-educated people, trying to keep students within the city borders after they graduate. Another strategy is to improve the supply of jobs for the higher educated, for instance by enhancing knowledge intensity in port-related activities connecting knowledge institutions with port-related business[3].

Tourism profile

In 2007, Rotterdam welcomed more than 1 million visitors who spent one or more nights in the city, more than ever before, but considerably less than the 8 million bed nights registered in Amsterdam (CBS, 2006 *cited in* OBR, 2006). Between 2002 and 2006, the share of Dutch hotel guests increased from 32 to 42 per cent. In absolute numbers, however, also more and more foreign guests arrive in the city: an increase of 23 per cent. Rotterdam Marketing (cited in OBR, 2007) reports that almost 25 per cent of the foreign visitors come from Great Britain. German tourists are responsible for 13 per cent of all bed nights, while visitors from the US take a share of eight per cent. In total, Rotterdam has 65 hotel accommodations, mainly in the three- and four-star category.

The total number of visits to Rotterdam– including so-called excursionists (same-day visitors) – has also increased: from 15.7 million in 2001 to 17.6 million in 2005[4]. Rotterdam has particularly been successful in attracting visitors to events such as the Rotterdam Filmfestival (358,000), the Rotterdam Marathon (825,000) and Rotterdam Summer Carnival (700,000). One of the most visited events is clearly related to industrial tourism: the World Port Days registers about 400,000 visits every year. The five most visited tourist attractions are the Zoo (Diergaarde Blijdorp, 1.5 million visits), the casino (874,000), subtropical swimming paradise Tropicana (478,000), Spido (375,000) and the Euromast (250,000) (OBR, 2006). Among the five most visited museums are the maritime museum (116,000 visits) and the port museum (94,500 visits).

Particularly relevant in view of our research are the activities of Spido. On its international website this company gives an overview of the tours that can be

3 This is the central aim of the Academic Centre TransPORT, a joint initiative of the cities and universities of Rotterdam and Delft, the Port of Rotterdam and Deltalinqs (representing port-related business).

4 This is the total number of visits. The number of visitors is lower as many people visit the city more than once.

Table 7.3 Visitors to Rotterdam

	2001	2002	2003	2004	2005
Bed nights	840,500	834,300	685,900	739,600	887,800
Visits (in millions)	15.7	15.3	17.3	16.4	17.6
Event visits (in millions)	4.3	3.6	4.7	4.3	5.7

Sources: Rotterdam Festival, Rotterdam Marketing, OBR (Kerncijfers Rotterdam, 2007; Plezierige Zaken in Rotterdam 2006)

booked. A standard harbour tour takes 75 minutes and is available all year through; in view of the limited time only a small part of the port can be visited. Recently, the attractiveness of this tour has been improved by investments in video screens that show things that can not be seen from the boat and provide additional information (e.g., inside photos of Hotel New York, explanations on how to load a container on a ship, etc.). Visitors that want to see the modern parts of the port (Europoort and Botlek) have to book extended harbour tours, with a duration up to 7 hours, which are only available during the summer (July and August). Furthermore Spido offers packages with traditional tourist attractions such as the Science Centre (Ontdekhoek), the Cube Houses, the Euromast and the Martime Museum, and one attraction that can be categorised as industrial tourism: shipyard De Delft (where a ship from the 18th century will be reconstructed). The combination of a harbour tour with a visit to this shipyard is, however, not very popular, with less than 300 participants a year.

3 The organisational framework for industrial tourism in Rotterdam

Tourism organisations

The organisation that is responsible for promoting Rotterdam as a tourist destination is Rotterdam Marketing. This organisation is the result of a merger in 2001 between several tourism organisations including the organisation responsible for tourist offices (VVV). Rotterdam Marketing wants to attract more tourists and business visitors, to increase average spending, and to enhance the image of the city as a destination. To that end, Rotterdam Marketing develops offers several services and products, and cooperates with stakeholders in the leisure industry. The foundation Rotterdam Marketing is funded by the OBR, the Rotterdam City Development Corporation, a municipal body responsible for promoting the city's economic development, and developing industrial areas, office locations and housing.

To improve coordination in city branding, the function of Chief Marketing Officer (CMO) was created in 2006. The CMO, who reports to the mayor directly and has some staff for support, has the task to promote coordination in city marketing activities, notably to improve the image of the city by ensuring consistent branding.

The Port of Rotterdam

The Port of Rotterdam plays a key role in developing and promoting Rotterdam's industrial tourism product. The Port of Rotterdam is the manager, operator and developer of Rotterdam's port and industrial area. Since 2004 the Port of Rotterdam is a public company of which the shares are owned by the municipality and the State.

According to a survey Dutch people consider the port the second most attractive element of Rotterdam together with shops (both mentioned by 31 per cent of the respondents) (TNS NIPO, 2006 *cited in* OBR, 2006). To meet the demand for visiting the port, several organisations – such as the Educational Information Centre Mainport Rotterdam, Rotterdam Marketing, World Port Days and Industrial Tourism – offer tours to the harbour targeting (leisure) tourists, companies and schools. In several of these organisations, the Port of Rotterdam has a stake. The Port of Rotterdam is also an important stakeholder in the development of a new tourist attraction, the Rotterdam Port Experience, and a visitor centre for the development of Maasvlakte 2 (FutureLand). Moreover, the Port Authority itself receives about 20,000 technical visitors a year.

The Educational Information Centre Mainport Rotterdam (EIC) The educational Information Centre Mainport Rotterdam (EIC) offers tours to the port and port-related companies. Although students and scholars are the centre's main target group, it also offers (tailor-made) packages to companies. In total, the Centre has a supply of about 70 excursions to a diverse range of firms in five categories:

1) Liquid bulk, including visits to several oil terminals (e.g., Vopak)
2) Dry bulk, with for instance tours to the EMO terminal
3) Container/general cargo, such as the ECT terminals
4) Industry, including tours to Akzo Nobel, ExxonMobil, Lyondell and Shell
5) Others, such as visits to Damen Shiprepair, Keppel Verolme, Mammoet and Smit Internationale

The EIC is located at Rozenburg right in the middle of the modern port area and comprehends an interactive exhibition of the region. The centre has strategic partnerships with companies and schools, and is financed by the City of Rotterdam and the Port of Rotterdam. It was set up by the Port of Rotterdam and Deltalinqs, an intermediary organisation that represents port-related companies. The main aim of the centre is to inform young people about jobs in the port and the courses offered by several educational institutions to gain the knowledge needed. In 2005, the EIC provided services to approximately 22,000 visitors (Deltalinqs, 2008).

Rotterdam Port Experience The Rotterdam Port Experience is a new tourist attraction, strategically located near the Erasmus Bridge and Spido, that opened its doors in October 2008. Initiators and financiers are the Port of Rotterdam and the City of Rotterdam, while Spido is in charge of the exploitation. The concept has been developed by Kranendonk Experience Network[5]. The Rotterdam Port Experience gives tourists and citizens the opportunity to see the port and the city, mainly by using new technologies and virtual reality in particular. The aim of the attraction is to inform people about the size, diversity and development of the port, and the relevance of the port for everyday life. Visitors take an elevator to the top floor of a virtual skyscraper from where they can see both the port and the city, gaining insight in the connection between the two. After that they make a virtual tour to the port, in a three dimensional virtual environment based on real recordings. Then they go to a 'container village' with computers in containers on which they have access to games and information. Finally they enter a room with a map of the city and the port on the floor and where different films can be seen. Here, they can also make a virtual cycling tour around the port. The total investment in this new attraction is €10 million, of which 80 per cent has been financed by the Port of Rotterdam. The exploitation is based on the expectation, supported by a survey-based market analysis, that the attraction will be visited by between 200 and 300 thousand people a year.

World Port Days Another organisation involved in arranging company visits is the foundation 'World Port Days' (Stichting Wereldhavendagen), which is responsible for organising the annual World Port Days. The event was invented by the port authorities in the late 1970s 'to bring the port back to the city', as port activities gradually moved away from the city along the Nieuwe Maas in the direction of the sea. Initially a department within the port authorities took care of the event, but later it was decided to set up a separate foundation, also in view of the fact that port authorities gained independence in 2004: in that year the Port of Rotterdam became a public company, owned by the local and national government. The foundation was set up to give it more freedom in acquiring funds, but also to avoid visitors holding the Port of Rotterdam liable for problems that occurred during the event. The Port of Rotterdam is, however, still the main financer of the World Port Days, covering approximately half of the budget while the municipal Rotterdam Development Corporation (OBR) finances about one-sixth. The remaining one-third of the budget is coughed up by participating organisations and companies. The foundation is directed by a board which consists of representatives from the Port of Rotterdam, which illustrates the strong connections between the two organisations that still exist. The Port of Rotterdam considers the event

5 Kranendonk Experience Network (www.kenworld.nl) has also been involved in the development of the Heineken Experience in Amsterdam and theme parks in the Netherlands and China.

increasingly as a strategic public relations tool, notably in view of labour market shortages and the increasing pressure on the license to operate.

In line with the event's mission (bringing the port back to the city), the main target group of the event is the regional population; about 65 per cent of the visitors live in Greater Rotterdam (which actually not only covers the administrative region, but also adjoining regions). Approximately 30 per cent originate from the rest of the Netherlands, leaving only 5 per cent for foreigners. Visitor statistics demonstrate that the event attracts a very diverse public in terms of age, ethnicity and education. The event is, hence, mainly oriented towards the Dutch audience, which is also reflected in the communication means (e.g., the website, brochures, and the language used during company visits). Professional visitors (e.g., suppliers and customers of firms) have not (yet) been identified as target groups; currently there are no special VIP arrangements. Through the event, the Port of Rotterdam (including the port-related companies and organisations) wants to secure its 'license to operate' and to support its image by showing the port's care for the environment and safety (to satisfy the needs of neighbouring residents and the population in general) and by gaining better access to the labour market. Although visitors (in this case defined as people living outside Greater Rotterdam) are not the main target group, they are considered increasingly important as the World Port Days have become a national and maybe even international image builder for the city. The event fits in the ambition of the city to be perceived as an attractive location in general, and for higher-educated knowledge workers in particular.

The World Port Days take place every year in September and last three days (Friday to Sunday). Each year a different theme is selected, such as 'pushing back frontiers in the world-port' (2006), 'working on a world-class port' (2007) and 'green light for the port' (2008). In the selection of themes the organisation responds to topical issues such as labour market shortages (2007) and climate change (2008). The event comprehends three types of activities: 1) presentations of port-related companies at the quay (near the Wilheminapier); 2) company visits (excursions); and 3) demonstrations of port authorities on the water, with fire fighter boats, crane boats, tow boats, etc. In view of our research we concentrate only on the second type of activities.

In total the World Port Days comprehends excursions to about 70 companies and organisations in different categories such as container transhipment, the environment (green), storage and loading, the petrochemical complex, maritime industry, energy, traffic control and food. In 2008 the programme also included four excursions that especially target young people (16–26 years). For many excursions an ID card is needed. On average, excursions take about three hours. For some company visits, age restrictions apply. Ticket prices vary between €5 for simple bus excursions to €49.50 for a boat trip including buffet. For most excursions the income from ticket sales does not cover the costs. Tickets can be reserved in the two weeks before the event takes place, by phone or through the internet. Many excursions are sold out quickly supporting the hypothesis that the demand for company visits is much higher than the supply. The total number of participants (including double counting for those visit more than one company) is approximately 25,000.

Many excursions take place more than once: for instance, the ECT terminal can be visited twelve times, including two special and very popular 'containers by night' tours. The programme not only includes visits to companies that produce consumer products – and are well known among people – such as Shell, BP, Q8 and E.On, but also to firms that mainly deliver to businesses such as LyondellBassell and Vopak. Apart from private companies also public companies such as the Port of Rotterdam and public authorities such as the Rijkswaterstaat (Department of Public Works) can be visited, as well as projects of the city (Stadshavens), and educational facilities (e.g., the test factory Oefenfabriek in Brielle and the Educational Information Centre Mainport Rotterdam). As part of the 2008 theme 'green light for the port', people could also choose to join one of the walking tours that show the beauty of nature in the port area. These walking tours fit in the Port's strategy to improve the physical appearance of the port by making the area more attractive and more accessible for visitors. This strategy has also resulted in the development of two cycle routes and some minor investments in green and visitor facilities.

Excursions are made either by bus, by train or by foot. Usually people are picked up near the quays of the Wilheminapier (where companies give their presentations), and on their way to the company they are going to visit, they already receive some general information on the port from the guide on behalf of the World Port Days foundation. In some cases this guide is also responsible for the company tour itself, but in other cases companies insist on using their own guides. The degree to which a company is involved in the visit differs from case to case: some firms arrange coffee and cake while others don't. Some firms want to use the event as a tool to improve their image, while others see it as a burden. One of the companies that never participated, despite its roots and headquarters in the city, is Unilever. The event organisers try to increase the involvement of companies by selecting themes that connect with their business interests. Particularly the labour market is a strong incentive: while the number of jobs in the port is growing (with the development of Maasvlakte 2 as one of the reasons), the supply of labour is decreasing because many companies have an ageing staff. Between 2002 and 2009, about 40 per cent of the staff is going to be retired (NRC Handelsblad, 2007).

The popularity of the World Port Days also made the Foundation think about the possibilities to open the doors of port-related companies on a more structural basis. In 2007, when the event celebrated its 30th anniversary, they experimented with one free excursion every month, in partnership with a publisher (Wegener). The (free) advertisements in several magazines and newspapers resulted in an average response of 250 per month, which was in line with the expectations of the publisher, but below the expectations of the organisers. For that reason the Foundation finds is difficult to judge if a year-round programme would be feasible. In any case, if industrial tourism will be organised on a more structural basis, these activities will compete with the three-day event.

Industrial Tourism Rotterdam

Between 1997 and 2003, the supply of industrial tourism was integrated in the municipal tourism organisations and subsidised by the Rotterdam Development Corporation: the VVV until 2001 and Rotterdam Marketing between 2001 and 2003. The initiative to offer visits to industrial companies was taken in 1997 by the current director of Industrial Tourism and the current Chief Marketing Officer who is also director of Cruise Port Rotterdam (an organisation that promotes Rotterdam as a cruiseship destination). Initially, the tours mainly targeted individual visitors: several buses departed from the Coolsingel (right in the city centre) to the various industrial attractions. Because the demand for these tours was not always sufficient, tours had to be cancelled often. From the very beginning Industrial Tourism had an international orientation, offering tours in various languages. For visitors with an educational motive, there was a clear task division between the VVV/Rotterdam Marketing and the Educational Information Centre (EIC) as the latter mainly targeted Dutch students. Then, in 2003, the Rotterdam Development Corporation and Rotterdam Marketing had to cut back on spending, with potential far reaching consequences for the industrial tourism activities. In response to this, one of the initiators decided to set up her own foundation (*Stichting Industrieel Toerisme*) which doesn't receive any municipal subsidies at all. With her knowledge of the market she decided to shift the target group of the tours from individual leisure visitors to groups of professional visitors.

The foundation Industrial Tourism is a very small organisation; the director takes care of marketing and promotion, and about 35 guides are hired and paid by the hour. Essentially, the foundation is a one-woman business. Industrial Tourism clearly has an international target group: the website is available in four languages (Dutch, English, German and French) and for tours even two additional languages (Spanish and Italian) are available. All guides are retired ex-workers of port-related companies: many firms insist on using their former employees as they want to have their story told in the best and most authentic way. Because they know many people on the work floor it is relatively easy for them to solve problems during a visit (e.g., when a bus of the foundation blocks a road, they solve this quick and in an informal way). In addition, visitors generally appreciate to hear personal anecdotes from the guide and from an organisational point of view, retired people are often more flexible than students and professional guides. For each tour, the director selects a guide that matches best with the specific profile of the group in question, also taking into account the availability and preferences of the guide.

The tours offered by Industrial Tourism are tailor made: based on the preferences of the customer the foundation compiles a programme with visits to one or more companies, if desired combined with visits to other attractions such as the Euromast or the Maritime Museum. The tour examples on the website (www.industrieeltoerisme.com) demonstrate that Industrial Tourism has a strong focus on harbour tours: obviously the Port of Rotterdam is the city's main industrial attraction, and one of the primary motives for international professional visitors to come to Rotterdam. All customers of Industrial Tourism are mainly interested in the port. The harbour tours, either by boat or by bus, are often combined with visits to specific companies such as the Rotterdam Car Centre, Seabrex (the fruit terminal) or the EMO terminal.

Industrial Tourism attracts about 10 to 15 thousand visitors a year, with an estimated share of foreigners of at least 40 per cent. Most of them are visitors with a business or educational motive. Many foreign delegations visit the port to see how it functions and to get inspiration for the redevelopment of ports in their own country. For the Dutch market, the foundation mainly targets staff associations (of public and private organisations) that want to organise an excursion that combines education with entertainment. Since Industrial Tourism is no longer subsidised by the Rotterdam Development Corporation, there is no division of work between Industrial Tourism and the Educational Information Centre concerning tours for students and scholars. Industrial Tourism also arranges tours for Dutch schools, while the Educational Information Centre also targets schools from abroad. Several guides work for both organisations.

In general visitors highly appreciate the tours provided by Industrial Tourism. One of the main challenges for the foundation is to make sure that expectations are met. Before the visit, people often expect that they can visit the port by boat only, but that would take too much time, with a distance of 45 km between the port and the city. For those want to see the port by boat there is a combined harbour tour by coach and by boat: from the Parkhaven, people go by boat to the Waalhaven and the Eemhaven (both older parts of the port), and then to the Geulhaven where they continue their tour by coach to the Europoort and the Maasvlakte. This implies that the most modern, and for professional visitors most interesting, parts of the port can only be visited by coach. A disadvantage of using a coach as mode of transport is the vulnerability for congestion on the roads between the city and the port: the ring road around Rotterdam is one of the most congested parts of the Dutch highway system.

A fast connection between the city and the port by boat or rail would definitely improve the accessibility of the product, although coaches will still be needed to visit the sites: most sites are huge in size and not accessible for pedestrians. During the company visit people have to stay inside the bus for security, health and environmental reasons, which to some extent limits the attractiveness of the visit. Moreover, at many sites it is forbidden to take pictures. Another limitation of the product is the lack of visitor facilities (e.g., restaurants, toilets, etc.), not only at the industrial sites, but also in the near surroundings. One of the few facilities in the port that is suitable for receiving large groups, is a hotel (De Beer). In general, companies in the port are willing to receive visitors as long as their reception does not conflict with safety, environmental protection, and the continuation of the primary process, but they are not (yet) willing to invest in special facilities for visitors or elements that could make the visit more attractive (e.g., information signs, exposition space). Port-related companies do realise that they have an interest in a more open relation with society, notably in view of labour market shortages, but their profit margins are too small to permit investments in industrial tourism.

4 Rotterdam's industrial tourism products

In this section we discuss in more detail five (potential) industrial tourism products in Rotterdam, based on interviews with representatives of the company in question. We discuss the activities of three companies in the food industry: Schmidt Zeevis (fish), Happy Shrimp Farm (shrimps) and Verstegen (spices). Smit Internationale and Broekman (car distribution) are representatives of the logistics and maritime sector. We want to emphasise that this is only a small selection of the companies that are open for visitors. In the previous section we showed that many port-related companies open their doors during the world port days, while some of them are accessible all year through via the foundation Industrial Tourism and/or the Educational Information Centre Mainport Rotterdam.

Schmidt Zeevis

Schmidt Zeevis is a Rotterdam-based company that processes, sells, and distributes fish and fish products to business customers all over the world, notably to hotels, restaurants, catering service providers and ship chandlers. The company, since 2000 a member of the Kennermervis Group, operates two stores, one in Rotterdam – at the same location where the founder once started – and one in Maastricht (in the south of the Netherlands). The 3,000 m^2 site in Rotterdam is located within a stone's throw from the river Nieuwe Maas and the Spido, but at considerable distance from the city centre. It consists of a shop with a small lunch corner and a site for distribution, storage and processing. The shop offers an impressive assortment of fish and fish products, unrivalled in the Netherlands, and is 'almost a tourist attraction in its own right' (Schmidt Zeevis, 2008, translated by the authors). Revenues account for almost 10 per cent of the overall turnover, but profits are relatively small. Schmidt Zeevis considers its shops as billboards for its core business: delivering products to business customers. Moreover, the shop is an instrument to get into contact with the end users, which is helpful to anticipate on their demands. Consumers are seen as important because they can influence the purchasers. Schmidt is medium-sized company: it employs around 120 people in total, including 14 who work in the shop.

Schmidt Zeevis is currently only open for professional visitors, with a business or educational motive. Every week the company receives on average two groups. They are mainly interested in the production process: the import, storage, processing and distribution of fish and fish products. Tours are given by the managing director, the commercial manager or one of the employees. A tour takes approximately 30 to 60 minutes and shows the visitor all elements of the production process. As you would expect from a food processing company, visitors have to clean their hands before they enter the work floor and they have to wear special plastic coats. The tour is made interactive by giving visitors the opportunity to ask questions and sometimes by allowing them to try filet fish themselves. Companies usually pay a small fee, while for most schools the tour is free of charge. The maximum group size is 15. There are no special facilities for visitors, apart from a toilet. Visitors need to be registered one day (24 hours) in advance, not for security reasons, but mainly because of planning (arranging a guide and in some cases a translator). While the shop is open from Monday to Saturday, the production site can only be visited from Monday to Friday.

In the view of Schmidt Zeevis the company tour is a relatively 'cheap', but also more effective and subtle form of marketing (e.g., in comparison with an advertisement in a specialist journal). Industrial tourism fits in the company's marketing strategy to develop relations with stakeholders, not only with primary stakeholders (customers, suppliers, financers, employees), but with secondary stakeholders (the general public, the media, etc.) as well. The main reason is that stakeholders are expected to influence one another. People who have visited the company can become ambassadors: they have seen how Schmidt Zeevis can guarantee the quality of its products (that is, by complying with several international standards[6]). The fact that several visitors also buy products in the store is an additional, but not the main reason for organising company tours. Another benefit mentioned by Schmidt Zeevis is the fact that company tours make employees feel proud about their job and the company they work for.

Schmidt Zeevis is willing to receive more visitors, and would also be interested in groups of leisure tourists, and not only because they buy products in the store. Particularly international guests could be of added value since they could play a role in further expanding the company's global network, reasoning that they might influence decision makers, or that they are decision makers themselves. An increase in the number of visitors requires, however, investments in visitor facilities (e.g., headphones, a presentation room) and professional guides (the director and his staff can not do all tours anymore). The question is if these investments outweigh the benefits, notably because most revenues are intangible and occur on the long run.

Happy Shrimp Farm[7]

In 2002 two employees of the Port of Rotterdam started to investigate the possibilities to use the waste heat of power plants for new sustainable activities nearby (co-siting). They were particularly interested in opportunities to grow fish with sustainable technology. After several years of research they selected the Pacific

6 Hazard Analysis Critical Control Point (HACCP) and British Retail Consortium (BRC).

7 The Happy Shrimp Farm went bankrupt in 2009. To keep this section readable it has not been adapated to this reality.

White Shrimp, raising the idea to develop a farm near the power plant of E.On Benelux, located on the Maasvlakte. In 2004 they started their own company: Bass & Gill (named after the founders Bas Greiner and Gilbert Curtessi). Happy Shrimp International, set up in December 2005, is one of the firms owned by this holding company; the adjective 'happy' refers to the sustainable and biological production process without only appealing to back-to-nature freaks. The Happy Shrimp Farm on the Maasvlakte covers an area of 5,000 m^2; it is connected with the power plant by a 2.5 km pipe, enabling the farm to use the waste heat of the power plant (Feed Tech, 2008). It was official opened in September 2006, during the 'Nigth of the Prawns' which had a second edition in September 2007. The initiators have signed an exclusive contract with Schmidt Zeevis (see previous sub-section) to distribute the relative expensive high-quality shrimps to top-class restaurants and catering companies that will put the Happy Shrimp on their menus (as a sort of brand). Together with E.On, Happy Shrimp International has the ambition to expand its business across Europe, starting with similar farms in Germany.

For the Happy Shrimp Farm, company visits are an integral part of a business strategy in which marketing, communication and public relations are crucial elements. From the very beginning the initiators have presented the project as an example of sustainable entrepreneurship, anticipating the growing concern about climate change. With clever communication, the Happy Shrimp Farm has won the sympathy of regional and national decision makers, resulting in strategic partnerships with public and private actors. On their website (www.happyshrimpfarm.nl), the company prefers to talk about 'friends' rather than partners or stakeholders. One of their friends is the Ministry of Economic Affairs that subsidised the initiative with €800,000 through the Senter Novem agency which promotes sustainable development and innovation. The Happy Shrimp Farm has succeeded to become one of the focal points of attention in the climate and energy debate, being an example of the 'Cradle to Cradle' philosophy (McDonough and Braungart, 2002), which has resulted in numerous requests for visits and presentations and massive media attention. The company's website includes a news section with more than 80 appearances in national and international media (newspapers, magazines, television).

Except during the World Port Days, when about 2,000 people enter the company, the site can only be visited by professional groups with a business or educational motive: customers, suppliers, financers, governments, schools, universities, research centres, etc. Company visits need to be beneficial for the company, either in terms of financial revenues (e.g., sales) or in terms of intangible benefits (e.g., reputation). In one year the company receives about 1,200 visitors; most of them from the Netherlands, but there are also visitors who come from other parts of Europe, China and South Africa. There are two types of visits. Every week the company is visited by two or three small delegations (up to five persons) with a clear business motive; because direct benefits are expected it makes sense that they don't pay for the tour. On average every month, the company gives a tour to a large group (30 to 40 persons) of people who are more interested in a combination of entertainment and education. An excursion to the Happy Shrimp Farm includes a presentation about the company, a workshop 'cooking and baking shrimps' and a lunch or dinner (NRC Handelsblad, 2008). In most cases they also get to see the basins where the shrimps (and also glasswort) are grown, although the company is a bit reluctant to allow visitors there (in view of safety and hygiene). To cover the costs of these tours, of which the benefits are more indirect, companies and ministries have to pay considerable amounts (up to €100 per person) for tours in combination with a workshop or training. Students and scholars can get an educational tour free of charge. The organisation of the visits has been contracted out to a private agency (Speakers Academy) which also takes care of requests for presentations.

In line with their strategy the initiators have from the very beginning expressed their ambition to open a visitor centre near the Happy Shrimp Farm. Together with E.On and Schmidt Zeevis, the Happy Shrimp Farm is investigating the possibilities, notably because the Farm itself is actually not very well equipped for receiving (many) visitors, also due to hygiene regulations. A visitor centre would take away the possibility 'to see the real thing' but it would enable more people (also residents and tourists) to learn about this best experience in sustainable entrepreneurship and to taste the prawns themselves. Not the product or production process is the primary attraction, but rather the whole philosophy behind the company. The visitor centre could act as a public relations and marketing tool for all actors involved. Moreover, it could include a restaurant that not only serves tourists but employees of nearby located companies as well. For tourists the location is well accessible by coach, car or boat, though at considerable distance from the city centre (45 km) making a conventional boat (normal speed) less suitable as mode of transport.

Verstegen

Verstegen is a family-owned company with a history of more than one hundred years (founded in 1893). The firm is a producer and supplier of herbs, spices, mixtures, sauces, marinades and various functional ingredients. In the Netherlands, Verstegen is a well known brand, visible in many supermarkets. Three business units target three main customer groups: consumers, professionals (wholesale to restaurants, institutions, caterers), and other firms in the food industry (that use the products of Verstegen as ingredients for their own products). Three other business units are more supportive, and take care of finance, purchasing and operations (technical assistance, security, logistics, etc.). The company's headquarters and production facilities are all located in Rotterdam, but the company sells its product to (business) customers all over the world. In Belgium and the UK, the company takes care of its own sales activities, while in Ireland they work together with a sales agent. There are plans to further expand sales activities in Europe.

In Rotterdam Verstegen employs around 360 people of whom 230 work in production, logistics and the technical division. The other 130 work in the headquarters and are responsible for sales, marketing and finance. Verstegen has three company buildings in Rotterdam, two of which are located next to each other in the Spaanse Polder, a business park near the A20 highway (in the north of Rotterdam). The third building, the logistics centre, is located in Rotterdam North West at short distance of Rotterdam Airport.

Initially, when the new production facilities in the Spaanse Polder were opened 12 years ago, Verstegen did not allow any visitors on its premises, with safety as mean reason. However, because the company also saw the benefits of company tours, they started to open their company step by step, anticipating the safety problems by introducing a screening procedure for guests (a questionnaire has to be filled in, and names and backgrounds of visitors are checked). Now, twelve years later, Verstegen regularly organises company tours, but still only for professional visitors with a business or educational motive. The main target groups are (potential) customers and educational institutions (schools and universities). The company approaches potential clients proactively and invites them for a company tour, showing them the quality of the products and the production process (e.g., hygiene). Every year Verstegen invites the last years' students of a vocational programme on food, but also other schools and universities are welcome. With two to three groups a week and an average group size of ten, the total number of visitors in one year is about 1,000. With an international customer base the firm also attracts international visitors: approximately 20 per cent of the visitors with a business motive come from abroad, notably from Belgium, France, Germany and England. Normally visits take place during the week and operation hours, although exceptions are made occasionally.

A typical company visit starts with a presentation and a short movie about the company followed by a guided tour which, on average, takes around 90 minutes. Before visitors enter the production facilities, they have to put on special clothing, a hair-net, glasses and overshoes, which is generally not considered as annoying, but rather as part of the experience. Company visits are tailored to the interests and the age of the group, but in principle all stages of production can be visited: from mixing sauces and bottling to deep-freezing. Only areas where particular patented processes or products are shown are not open to everybody (e.g., competitors). Other areas can only be seen from behind glass plates, due to hygiene standards (e.g., the production of sauces). And if a group of kids visit the factory, moving machines that could be dangerous for them are avoided. During the tour visitors can ask questions and talk to production workers, as long as safety procedures are not violated. At the end of the tour each visitor gets a little present: a box that contains a selection of Verstegen's products. Visitors are guided by employees in higher or middle positions; with 22 nationalities represented in the company's staff the firm can offer tours in many languages, but particularly in Dutch, English, German, French and Spanish, the same five languages that can be selected on the corporate website (www.verstegen.nl). Employees that want to act as guides have to follow an internal training in advance. For larger groups (with more than 15 visitors) the company even arranges two guides in order to secure safety.

Verstegen's company tours are not (yet) available for consumers (leisure visitors), although they get many requests. Currently, the main reason for not opening its site for consumers is that safety cannot be guaranteed when more visitors arrive: if leisure visitors are to be allowed, professional guides have to be hired which would require investments of which the revenues are uncertain (would the firm attract enough visitors to cover the costs?). Consumers are considered important for the reputation of the firm, but in comparison with business customers and educational institutions they are considered less important (but not unimportant!) as a stakeholder. With the limited capacity to receive visitors the company targets the two groups that are most relevant.

Smit Internationale

Smit Internationale was founded in Rotterdam in 1842. Today it is a global player in the maritime services sector, still headquarted in Rotterdam but with about 50 locations worldwide. The company has divided its activities among four divisions: Harbour Towage (harbour towage and related maritime services), Terminals (towage services and related maritime and management services to offshore and onshore terminals), Salvage (salvage, wreck removal, environmental protection and consultancy) and Transport and Heavy Lift (chartering, barge rental and transport, ocean, coastal and river towage, heavy lifting and marine support to a variety of civil and offshore projects). Smit Internationale gives employment to about 4,500 employees and an additional 1,500 in joint ventures. In Rotterdam, the company provides jobs to approximately 250 people, including many higher technical and management staff members. In contrast with the perception of many, the maritime services sector is very knowledge intensive. Smit Internationale got international media attention with the salvage of the Koersk, together with Mammoet, a Dutch company headquartered in Schiedam and specialised in heavy lifting and transport. Together they published a documentary with the title 'Raising the Kursk' (2002), which was sold to several television stations including Discovery.

Smit Internationale regularly receives technical visitors: once or twice a week the company gives a tour to groups with a clear business or educational motive. Because the demand for visits is much higher than the supply (the capacity is limited), the firm has to be selective in honouring requests. Primary stakeholders – actors who are closer to the primary process – clearly take precedence over secondary stakeholders. One of the major attractions at the Rotterdam location is a warehouse where equipment for salvage operations is stored (three similar facilities are located in other parts of the world). The tugs are also interesting, but these vessels cannot be found near the headquarters but rather at strategic locations in the region or on the sea. Apart from some small vessels, there is not that much to see in the area around the company's main office. Visitor facilities are present though: a floating platform has been developed into the ADMI society: a place for receptions and (video) presentations, with a maximum capacity of 50.

The company is not open for leisure visitors. Smit Internationale cannot be visited during the World Port Days, although the company does recognise its interest in building sustainable relations with its environment. Safety and security reasons explain to a great extent why the doors remain closed for many people. The equipment for salvage operations in the warehouse is not only very interesting, but also very expensive. For these and other reasons the company has decided to concentrate on other forms of participation in the World Port Days: the firm gives presentations at the quay, and is involved in the organisation of a mini seminar at a boat, together with several other organisations/companies. Smit Internationale is not only involved in this event because of public relations, but also to gain better access to the labour market. The company's Human Resources department is closely involved in these kind of initiatives.

Broekman Automotive

Broekman Automotive is one of the five divisions that constitute the Broekman Group. Broekman was founded in Rotterdam, in 1960, and has developed into a global company with locations in Belgium, Hungary, the Czech Republic, India and Singapore, providing employment to about 1,100 people in total, of whom 200 work in Rotterdam. The automotive division consists of five companies in Rotterdam and one in Born (Limburg). One of these companies is the Rotterdam Car Centre which is one of the industrial tourism attractions offered by the foundation Industrial Tourism. This site covers an area with a storage capacity of about 30,000 cars. The Rotterdam Car Centre offers a wide range of services in addition to storage, like damage inspection and repair, dewaxing and polishing, and even the fitting of small modifications such as sunroofs.

The Rotterdam Car Centre has in general a positive attitude toward the reception of visitors. Through the foundation Industrial Tourism they receive groups on a regular basis. However, because these groups do not necessarily correspond with the target groups of Broekman (the company is not involved in selecting the visitors), the company does take a somewhat defensive approach, e.g., by restricting the number of visits per week. The bus of Industrial Tourism has to arrive between the workers' morning break and their lunch time, enabling the use of the canteen as reception room (in absence of special visitor facilities). Moreover, the Car Centre only allows former (retired) employees who now work for Industrial Tourism to do the tour; this is mainly because former workers have the knowledge needed. Next to the tours via Industrial Tourism, Broekman also receives groups that contact the company directly. In that case the company is more selective, only accepting requests by organisations that are somehow linked to the company's business. In any case it is not allowed to take pictures.

During a visit to the Rotterdam Car Centre visitors get to see various workshops. In contrast with car factories it is also possible to see the paint shop because it deals with repairs rather than the initial coating of a car. Another attraction is the real time location system based on radio frequency identification technology which continuously shows the location of each car. The Car Centre receives many other companies and scientists who are interested in this system.

The company is involved in industrial tourism for two reasons: one is to improve its reputation, another is to gain better access to the regional labour market. This last reason is becoming increasingly important as many vacancies cannot be filled. Another argument for the company to be involved in industrial tourism is the relation with the city: Broekman wants to be a good citizen.

5 Applying our frame of analysis

In this section we apply our frame of analysis to the observations made in Rotterdam. We discuss the following issues: 1) adequate target group selection; 2) the attractiveness of the industrial tourism product); 3) benefits in terms of income and image for city and enterprise; and 4) the organisational setup.

Adequate target group selection

The City of Rotterdam has no explicit industrial tourism strategy, so no official target-group selection is available. In general, however, it can be said that for both leisure and business visitors the port is a crucial element, which is also reflected in the new tagline 'world port – world city'; this motto has been defined as part of an international branding strategy under the supervision of the Chief Marketing Officer which targets international businesses and investors. The foundation Industrial Tourism, the Educational Information Centre, Spido and the event World Port Days all target visitors who are mainly interested in the port. The port is clearly a primary tourist attraction for the city.

Another observation is that business groups receive more attention from suppliers of industrial tourist attractions than individuals (residents and tourists). Individuals (citizens and tourists) that want to see the port cannot book tours at the Educational Information Centre or the foundation Industrial Tourism. The fact that Industrial Tourism decided to focus on business customers after it became a private, non-subsidised foundation tells something about the demand for company tours: the demand from business and educational groups is estimated higher than the demand from individuals. Individuals can, however, visit the port with Spido, but they can't see the port-related companies. The World Port Days attract many individuals (25,000) – proving that there *is* considerable demand from individuals – but this annual event only lasts for three days. Also companies seem to prefer business groups. Looking at the scope of the visitors groups targeted by the companies that we analysed, we can observe that two of the four companies in our sample only focus on their primary stakeholders.

Verstegen, a consumer-oriented company, is the most restrictive firm, only allowing business customers and students to enter its premises (some years ago, even these groups could not visit the company). Because of limited capacity, other primary stakeholders such as suppliers and consumers are not able to visit the site. Smit Internationale, a business-to-business company, only receives visitors who are considered relevant in view of the primary production process (customers, suppliers, financers, etc.). The Happy Shrimp Farm and Schmidt Zeevis are both examples of companies that take a somewhat more proactive approach to industrial tourism, targeting a broader range of stakeholders. Both companies not only target primary stakeholder groups, but secondary stakeholders as well. They can also be visited by groups who are not directly interested in doing business. The Happy Shrimp Farm is particularly active in attracting government officials and the media – who can influence other stakeholders – as part of its public relations strategy. Schmidt Zeevis argues that any contact with the outside world can be beneficial for the company through indirect relations, although influential decision makers and higher-educated people in general are clearly seen as more powerful in this respect. Although both companies are consumer-oriented (they produce consumer products) consumers are not (yet) recognised as target groups of company visits (a visit to the shop of Schmidt Zeevis cannot be seen as industrial tourism). Broekman Automotive is a business-to-business company, but clearly deals with consumer products. Normally the firm only accepts delegations that are relevant for the company's business, but through the foundation Industrial Tourism they receive groups with very different profiles, not necessarily linked to the company's business.

For companies the scale of the target groups is closely related to the markets on which they are active. All companies that we analysed operate on a global level, so it is no surprise that business visitors come from many places, although the majority come from the Netherlands. Educational groups (schools, universities) mainly come from the region where the companies recruit most of their staff. This is also illustrated by the fact that the website of the Educational Information Centre is in Dutch only. Only for the Happy Shrimp Farm the region is somewhat less relevant as supplier of labour since the production of shrimps requires highly specialised staff, recruited from all over the world. For companies in the port and industrial cluster the regional labour market is gaining importance as motive to open their doors to society, notably in view of high average age of their staff (many workers will retire in the years to come). In line with this the World Port Days mainly focus on visitors from the region, not only because they are potential employees but also since citizens are interested in a clean and safe port. Although the World Port Days have the potential to attract international visitors, the website is only in Dutch, which confirms the limited scale of the target group. The foundation Industrial Tourism has a more international focus with a relatively high share of foreign visitors and a multilingual website.

The attractiveness of the industrial tourism product

The port of Rotterdam is an attraction in its own right, with several highlights such as the ECT and EMO terminals, the Rotterdam Car Centre and the Fruit Terminal

(Seabrex). Particularly the size of the port – and of nearly everything that can be found in the port – is impressive. The port's size is, however, also a factor that makes the port less attractive in terms of internal and external accessibility. With conventional ships it takes a lot of time to travel from the city to the Maasvlakte (1, 2) and Europoort, and also within the port area there is only one mode of transport available: the coach. Most things can only be seen from behind the windows of the bus, as walking is not an option not only because distances are too long but also in view of safety and security regulations. The foundation Industrial Tourism admits that the distance between the city and the port is a barrier, and a potential source of 'unattractiveness', also in view of the congestion on the roads between the city and the port. To secure customer satisfaction the foundation's director makes sure that the expectations of the visitors are in line with reality. Another instrument to improve the attractiveness of the product is the development of packages: Industrial Tourism offers standard and tailor-made combinations of company visits, but can also assist in combining company visits with visits to more traditional tourist attractions. We have also seen that the modern port area has a disadvantage because of a lack of facilities.

Particularly for professional visitors the attractiveness of an industrial tourism product does not depend too much on the quality of visitor facilities and the appeal of the product and the production process. Professional visitors are mainly interested in innovative production and management methods. This is what could be learned from the experiences of the Happy Shrimp Farm: as an example of sustainable entrepreneurship this company has become very attractive and inspiring for business visitors with educational motives. In line with this, a visit to Schmidt Zeevis is particularly interesting because it informs the visitors about quality control and the logistics process. Interestingly, both the Happy Shrimp Farm and Schmidt Zeevis make the visit more attractive by allowing visitors to prepare food themselves. All three companies in the food sector (including Verstegen) pay a lot of attention to hygiene, but this is generally not considered as a problem but rather as an element that adds to the experience of visiting the company. In the case of Smit Internationale the main attraction is a story, the salvage of the Kursk, which has become a symbol for the company's expertise; a story which becomes more attractive if combined with a visit to the warehouse where salvage materials are stored. The Rotterdam Car Centre is one of the few companies in Rotterdam that deals with appealing consumer products: cars. The Car Centre demonstrates that also logistics centres can be attractive for visitors (both leisure and business), notably because these kind of facilities also make small modifications. Strikingly, the paint shop of the Rotterdam Car Centre can be visited while this is not possible at many car factories.

Although the Rotterdam Port Experience is not an industrial tourism product, according to our definition, it can help to promote the attractiveness and economic relevance of the port. By enabling visitors to enter the port in a virtual way they make become more interested in visiting the port and port-related companies in reality. The Rotterdam Port Experience can thus act as a portal for industrial tourism, increasing the accessibility of company visits.

Benefits in terms of income and image for city and enterprise

The companies we analysed expect to receive several benefits in return for their investments in industrial tourism. All companies recognise industrial tourism as a tool of stakeholder management, but they seem to have different views on the scope of the group to be targeted. It appears that proactive companies (Schmidt Zeevis and the Happy Shrimp Farm) expect more benefits from receiving secondary or indirect stakeholders than more defensive companies (Smit Internationale, Verstegen, Broekman) do. Other companies in the port, not in our sample, probably also take a more defensive approach. They mainly expect benefits from industrial tourism through enhanced access to the labour market. Smit Internationale and Broekman, however, are also involved in industrial tourism – receiving citizens and tourists through Industrial Tourism and the World Port Days – because they want to be good citizens and they want to secure their reputation. In our sample Schmidt Zeevis and the Happy Shrimp Farm are the only companies that benefit from industrial tourism directly. Schmidt Zeevis earns some extra money by selling fish and fish products in its shop (which is open for everybody, but also to people who visited the company), while the Happy Shrimp Farm gains income through selling workshops in combination with a company visit/presentation. Both companies, however, use the profits they make on business visitors to subsidise visits from schools. Schmidt Zeevis is the only company in our sample that explicitly mentioned the fact that company visits make employees feel proud, another benefit which can contribute to productivity, and hence, to profitability.

The City of Rotterdam has no explicit industrial tourism policy. However, because the municipality subsidises the Educational Information Centre, it can be stated that the City expects some benefits in return: this centre helps to promote the reputation of the port (and industrial cluster) as an employer. Also the €2 million subsidy to the development of the Rotterdam Port Experience fits in the strategy to improve the image of the port, to make (young) citizens and visitors more aware of the economic relevance of the port and to demonstrate the knowledge intensity of the port. Currently, industrial tourism is not considered an interesting niche market for attracting visitors. In 2002 subsidies to industrial tourism were cancelled, also in view of the disappointing demand from leisure tourists. After that the foundation Industrial Tourism shifted its attention to business visitors, but without the financial and political support of the municipality. Nevertheless, it could be said that the activities of Industrial Tourism add to the image of the city and the port in the same way as the Educational Information Centre does, and probably even more to the international image in view of its customer base. With 10 to 15 thousand visitors a year, Industrial Tourism can also be expected to generate additional income for the tourism industry. This may not always apply to international visitors – who would come to the city anyway and use the tours to see the city and the port – but it may be true for the national business and school excursion market, particularly because port visits are combined with visits to traditional attractions and a lunch or dinner in the city.

The organisational set-up

The way industrial tourism is organised within a company seems to depend on the objectives. For instance, at Smit Internationale, we have seen that the Human Resources department is highly involved in the activities during the World Port Days, as this event is used as a tool for recruitment. At Verstegen, which has a very defensive approach, the business unit operations is mainly responsible for organising visits: it is the task of this department to make sure that visits do not conflict with the day-to-day operations and security in particular. The Happy Shrimp Farm and Schmidt Zeevis consider company visits as an essential element of their public relations strategy. In absence of a public relations department, the managing directors of both companies take a leading role in the reception of guests.

On the level of the city we observe that there are several organisations somehow involved in industrial tourism. Strikingly, two organisations – Industrial Tourism and the Educational Information Centre – are more or less competing for the same visitors, while some guides work for both organisations. Both organisations have a strong link with the Port of Rotterdam and the foundation World Port Days in particular. As far as tourism promotion is concerned, Rotterdam Marketing plays an important role. All organisations claim there own role, and the question is how to coordinate all this. This automatically brings us to the Chief Marketing Officer (CMO) who has been installed to improve coordination in the city's marketing activities. If Rotterdam decides to invest in developing and promoting industrial tourism, you would expect the CMO to make sure that these investments fit in the overall branding strategy.

References

COS 2007. *Kerncijfers Rotterdam 2007*. Rotterdam: COS.

Deltalinqs 2008. [Online]. Available at: http://www.deltalinqs.nl [accessed: 2008].

ECT 2008. [Online]. Available at: http://www.ect.nl [accessed: 2008].

Industrial Tourism Rotterdam 2008. [Online]. Available at: http://www.industrieeltoerisme.com [accessed: 2008].

Koeleman, E. 2008. Sustainability drives growth Happy Shrimp Farm. *Feed Tech*, 12-1-2008, pp. 18–20. [Online]. Available at: http://www.agriworld.nl/public/file/pdf/20080213-18_fet_happyshrimps.pdf [accessed: 2008].

McDonough, W. and Braungart, M. 2002. *Cradle to Cradle: Remaking the Way We Make Things*. New York: North Point Press.

NRC Handelsblad 2007. Vastgeroest beeld van vies werk breekt haven op. *NRC Handelsblad*, 10-09-2007.
NRC Handelsblad 2008. Werken, werken en nauwelijks iets te zien. *NRC Handelsblad*, 27-03-2008.
OBR 2006. *Plezierige Zaken in Rotterdam; Vrijetijdseconomie in cijfers*. Rotterdam: OBR.
OBR 2007. *Economisch Verkenning Rotterdam 2007*. Rotterdam: OBR.
Port of Rotterdam 2008. [Online]. Available at: http://www.portofrotterdam.com [accessed: 2008].
Rotterdam Airport 2008. [Online]. Available at: http://www.rotterdam-airport.nl [accessed: 2008].
Schmidt Zeevis 2008. [Online]. Available at: http://www.schmidtzeevis.nl [accessed: 2008].
Shell Pernis 2008. [Online]. Available at: http://www.shellpernis.nl [accessed: 2008].
World Port Days 2008. [Online]. Available at: http://www.wereldhavendagen.nl [accessed: 2008].

Discussion partners

Ms. Claudia van Andel, Public Relations Manager, Smit Nederland.
Mr. Marcel van Breda, Schmidt Zeevis.
Mr. Gilbert Curtessi, Happy Shrimp Farm.
Ms. Mai Elmar, Director Cruise Port Rotterdam, Chief Marketing Officer.
Ms. Annie Gruisen, General director, Spido.
Mr. Rene Hanssens, OBR (City Development Corporation), Rotterdam.
Ms. Corinne van Iersel, Port of Rotterdam.
Mr. Jan Kerremans, Director Operations, Verstegen.
Ms. Marianne Klein, Bosscher & Klein former director of Rotterdam Marketing.
Ms. Annemieke Loef, World Port Days (Stichting Wereldhavendagen).
Mr. Frits Mehrtens, Managing Director, Broekman Automotive Divison.
Ms. Maartje den Teuling, Chamber of Commerce Rotterdam.
Mr. Kees Vrijdag, Chamber of Commerce Rotterdam.
Mr. Ton Wesselink, Director Rotterdam Marketing.
Ms. Irma de Winter, Stichting Industrieel Toerisme.

Chapter 8
Conclusions and Recommendations

1 Introduction

In view of the increasing demand for company tours and the growing need for companies to open their doors to society (securing their 'license to operate', being a good corporate citizen) expectations are that industrial tourism can be beneficial for both city and enterprise. Company visits can generate additional income to the benefit of all, and in addition they can help to improve the image of a city and the industries located there. In practice, however, several barriers seem to complicate the development of industrial tourism, such as safety issues and the lack of visitor facilities. Therefore, we came to the following question as guideline for the present study:

Under what circumstances can industrial tourism make a strategic contribution to the competitiveness of city and enterprise?

In this chapter we translate the observations in the cities and regions we analysed to general conclusions and recommendations. The current chapter largely follows the structure of the frame of analysis that we developed in Chapter 1, which identifies four categories of conditions that are relevant in the development of industrial tourism: adequate target group selection, an attractive product, benefits in terms of image and identity for city and enterprise, and the organisational setup. First, however, we address two fundamental questions: what is industrial tourism and is there demand for industrial tourism? In the final section we present our recommendations.

2 What is industrial tourism?

In the Netherlands 'industrial tourism' (*industrieel toerisme*) has developed into a common notion for visiting operational firms. Both Rotterdam and Amsterdam seem to have discovered industrial tourism as an interesting segment of the tourism industry to develop[1]. In the international tourism literature, however, we found that industrial tourism – in a translated version or not – has many definitions, sometimes also encompassing visits to non-operational firms, while at

1 See www.industrieeltoerisme.com (Rotterdam) and www.industrieeltoerisme.nl (Amsterdam).

the same time many other notions are used to indicate visits to operational firms, such as technical visits. Some see industrial tourism as a sub-group of 'economic discoveries' (together with industrial heritage and science museums), while others consider company visits as a sub-group of industrial tourism.

In the first chapter we defined industrial tourism as visits to operational firms (not only manufacturing industries) of people (not only tourists) who are interested in the firm, its products, the production process, and/or the way a company is managed. We admit, however, that the term industrial tourism is a bit misleading as it is often associated with industry (as in *manufacturing industry*) and tourism (as in *leisure tourism*). This particularly applies to the Dutch audience as the term industry (*industrie*) is hardly understood as a line of business (*branche*), thus excluding visits to non-manufacturing firms.

The confusion of tongues concerning industrial tourism also manifests itself in the case studies:

- In the case of **Wolfsburg** we must conclude that Autostadt is actually an industrial theme park, or *brands park* (for all brands of the Volkswagen Audi Group), often seen as a good example of industrial tourism, but not exactly an example of a visit to an operational firm. The visit to the operational Volkswagen factory is officially *not* an integrated part of Autostadt although many people visit both. The theme park and the factory meet each other in the power plant which is partly operational (for the factory) and partly used by Autostadt for events (as industrial heritage).
- In **Cologne** both the tourist board and the chamber of commerce prefer to use the more general term 'company visits' (*Firmenbesichtigungen*) or the more specific term 'technical visits', but not 'industrial tourism' which is most associated with industrial heritage tourism; company visits have also been integrated in the event 'Expedition Colonia' which provides opportunities to discover the city. Apart from several company visits (to UPS, Rheinenergie, Magic Media Company, etc.) we also found an example of a brand park: BayKomm, the communication centre of Bayer. Interestingly, also a visit to a waterfront redevelopment project (Rheinauhafen) is in the list of technical visits.
- In **Pays de la Loire** regional tourism policy makers talk about 'economic discoveries' (*découvertes économiques*) as a group name for visits to 1) companies; 2) industrial heritage; and 3) science museums. The regional association Visit Our Companies (*Visitez nos entreprises en Pays de la Loire*) wants to contribute to the regional policy to develop this sector, mainly by promoting the development of the first sub-group, also referred to as 'industrial tourism' (*tourisme industriel*) and 'company visits' (*visites d'entreprises*). The association can only be joined by *active* companies; but no matter in what sector they are active (manufacturing, agricultural, services, etc.). Interestingly, also visits to so-called display sites (*sites vitrines*) are seen as company visits, suggesting that also corporate

communication centres, company museums, factory outlets and company theme parks could be promoted by Visit Our Companies. Our sample of examples in this region not only includes visits to factories (Airbus, Aker Yards, Cointreau) but also to the Port of St Nazaire, the salt marshes of Guérande and several research centres in the field of horticulture.

- In **Turin** they developed a programme called 'Made in Torino; Tour the Excellent' which comprehends visits to *active* companies only, although not necessarily to their production sites. Near their production sites which cannot be visited, car design companies such as Pininfarina and Giugiaro have set up display sites and museums which *can* be visited. Among the production sites that can be visited are the factories of Fiat, Aurora (pens) and Gufram (design objects and furniture). On a regional level, industrial tourism is mostly associated with industrial heritage (such as Lingotto, the old Fiat site in Turin) and agricultural tourism (e.g., visits to vineyards).
- **Shanghai**'s Industrial Tourism Promotion Centre promotes the development of six categories of industrial tourism – they use this term in English – including individual factories, business parks, museums and exhibition centres, and even industrial land marks such as the Maglev high-speed train line. The centre clearly uses a rather broad definition of industrial tourism, not only approaching manufacturing firms, but also service companies: participation of service companies is considered to be crucial, as Shanghai wants to be perceived not only as a place where manufacturing takes place, but also as a creative place. In this case study we analysed visits to individual factories of companies such as Baosteel, Shanghai Volkswagen and Yakult, but also visits to one creative business park (M50).
- Many scholars and policy makers refer to **Rotterdam** as one of the places where industrial tourism was developed and promoted first. From the very beginning, however, industrial tourism has been mainly restricted to visiting the port and its related industry, taking the activities of the foundation Industrial Tourism, the Educational Information Centre and the World Port Days into consideration. Also the companies in our sample – Broekman, Happy Shrimp Farm, Schmidt Zeevis, Verstegen, Smit Internationale – belong to the port and industrial complex. Only very recently, the Chamber of Commerce put 'technical visits' on the agenda, suggesting that also companies from other strong sectors in Rotterdam (creative industries, medical industry, services, etc.) could open their doors for visitors, though with a clear emphasis on visitors with an educational and business motive.

We conclude that the six cities/regions that we investigated use different definitions of industrial tourism and different names for industrial tourism. Definitions of industrial tourism range from very narrow (Rotterdam) to very broad (Shanghai), which makes it difficult to compare activities and their results, for instance in terms in visitor numbers. No matter what definition is used, it is important to understand the position of company visits in relation to other types of 'economic discoveries' and other tourist attractions. For policy makers, it seems wise to take a somewhat broader view, approaching company visits as a sub-group of economic discoveries or industrial tourism in the broadest sense of the word, paying attention to potential synergies with other sub-groups such as industrial heritage, science museums and company theme parks.

3 Is there demand for industrial tourism?

Several studies indicate that there is a growing demand for industrial tourism. When we started our research we expected the demand for industrial tourism to grow, notably in view of the increasing demand for 'purposeful visits', combining pleasure with education. But what can we learn from the case studies?

- In the exceptional case of company town **Wolfsburg**, visits to Volkswagen generate the majority of the overnight stays, while 90 per cent of the package trips to the city include a visit to Autostadt. The park attracts 2 million visitors a year, while the factory of Volkswagen receives almost 200,000 people a year and even more requests (the capacity is limited).
- In **Cologne** the Tourist Board considers industrial tourism a 'niche segment' of the tourism industry. Nevertheless, the companies we analysed, report substantial visitor numbers: 6,200 for Rheinenergie, 10,000 for the Magic Media Company and almost 120,000 for Bayer's communication centre (average per year).
- In **Pays de la Loire** the growing demand for economic discoveries was one of the occasions for the region of Pays de la Loire to set up the association Visit Our Companies: the initiators expected that people are increasingly interested in their roots, in the working life of their fathers and forefathers, and in the economic identity of the region they live in or visit. In 2004, company visits took a share of 11 per cent in the total number of visits to regional tourism attractions, corresponding with 1 million visitors a year. The Port of St Nazaire and Terre de Sel both report increasing visitor numbers (attracting 10,000 and 55,000 visitors respectively), while Cointreau and Aker Yards experienced a decline recently (to 30,000 and 22,000 respectively), but this has to do with very specific reasons (an increase in the entrance fee at Cointreau and the fact that the number of visits to Aker Yards very much depends on the cruise ships under construction).
- For **Turin** and the Piedmont region industrial tourism is a small segment of the tourism industry, at least according to the policy makers. There is, however, considerable demand for company visits. Fiat receives on average 23,000 visitors, while Aurora registers more than 2,000 visitors a year. Moreover, we observed that the supply generated by the Made in Torino Tour is insufficient to meet the demand from not only inhabitants but tourists as well.

- In **Shanghai** industrial tourism has been recognised as one of the ten key segments of the tourism industry, with 6 million visitors in 2006 (a share of 6.4 per cent of all visits to attractions), compared to an estimated 2 million in 2004. It has to be said however that the usefulness of these data is questionable, also in view of the broad interpretation of industrial tourism. Nevertheless, it can be said that the demand for industrial tourism has increased for sure, notably because domestic tourism in China experienced an impressive increase of more than 100 per cent in the last four years. Individual companies are successful in attracting large numbers of visitors, with 80,000 visits to Baosteel, 100,000 to Shanghai Volkswagen, about 12,000 to M50 and almost 70,000 to Yakult.
- In **Rotterdam** the foundation Industrial Tourism and the Educational Information Centre together arrange tours for about 35,000 customers a year. The company tours during the World Port Days generate another 25,000 company visits, and expectations are that the Rotterdam Port Experience, not exactly a company visit but rather a brand park for the port, will attract up to 300,000 visitors a year. In addition, individual companies also attract substantial numbers of technical visitors, ranging from about 1,000 for the Happy Shrimp Farm and Verstegen to 20,000 for the Port Authority (annually). During the year the demand for company tours from individual visitors (citizens and tourists) seems considerably lower and less stable, at least if we take into consideration that the foundation Industrial Tourism decided to focus on business visitors when it became independent from the municipality in 2003.

The general conclusion is that there is a substantial demand for industrial tourism. There are many indications that the demand for industrial tourism is growing, supported by the fact that several actors invest in industrial tourism because they expect the market to grow. Obviously, for cities the demand for 'economic discoveries' very much depends on the kind of attractions within their borders. Big companies with well-known names such as Volkswagen, Airbus, Baosteel and Fiat are able to attract more visitors than the smaller firms. Furthermore we emphasise that the size of the industrial tourism market very much depends on the definition of industrial tourism, since industrial museums and industrial theme parks can attract many more visitors than factories can. Another factor of importance is the size of the local market, partly explaining the impressive visitor numbers in Shanghai.

Visitor numbers give some idea about the demand for industrial tourism, but more information is needed to get the full picture. Aggregated data on industrial tourism is not available in all cities/regions, making it hard to draw any conclusions and to make comparisons between the case studies. Therefore, it has not been the aim of our study to actually *prove* that the demand for industrial tourism is growing, because to answer such a question, a more in-depth quantitative analysis would have been needed. Moreover, we emphasise that visitor numbers are restricted by the capacity of companies to receive visitors: in our view the actual demand for company tours can only be measured by means of market research. And finally, not only quantitative but also qualitative information concerning the demand is needed: what groups in society are actually interested in visiting firms? As we will see in the next sections, the quality of the demand is often more important than the quantity.

4 What groups should be targeted? And what groups are reached?

Industrial tourism may target several groups, ranging from leisure tourists to students, professionals, journalists and researchers. The case studies provide us with insights on what groups are targeted by the actors involved, but also on what groups are actually reached. What are the main observations and conclusions?

- The experiences of Volkswagen (**Wolfsburg**) make it clear that for companies it makes sense to target specific groups, without excluding groups. Because the factory tour is offered to people who pick up their car, they automatically reach one of the most important target groups: customers. Another effective strategy to reach particular target groups is to offer special (VIP) tours for them. However, as a consumer-oriented company Volkswagen is happy with any visitor, for the simple reason that every visitor is a potential car buyer or at least involved in the decision of a car buyer (e.g., children). The share of international visitors to Autostadt is 9 per cent, and an additional 51 per cent travel more than 100 km, illustrating the large catchment area of an industrial theme park.
- In the **Cologne** region, Bayer (Leverkusen) targets specific groups of visitors, trying to reach opinion makers with considerable influence on other actors. Other companies focus on one specific stakeholder group, such as customers (Rheinenergie) or universities and schools (Knapsack). The Chamber of Commerce and Industry has identified foreign investors and business visitors (participants of conferences and trade fairs) as potential target groups of company visits. Leisure tourists appear to be less interesting as a target group: many tourists (not only from Germany but also Austria, where German TV shows are broadcasted) are interested in the media sector, but media companies such as Magic Media seem to adopt a defensive approach (visitors will come anyway, there is no need for active promotion). Energy company Rheinenergie found a very simple solution for allowing individuals to participate in tours: they can join groups (which is possible because they all participate for free).
- The **Pays de la Loire** case shows that companies, in general, are more interested in attracting primary stakeholders (customers, suppliers, shareholders, employees) than the general public (secondary stakeholders, including citizens and tourists). Particularly young people are targeted, in

some cases because they are potential employees (e.g., Terre de Sciences), in other cases because young people are desired as consumers (Cointreau). Visit Our Companies has decided not to define target groups on a regional level, leaving this task to companies and travel agencies. The association does assist, however, by providing visitor statistics and promoting the product nationally and internationally, in cooperation with the regional tourism bureau. The average share of international visitors is high, with 25 per cent. Particularly British tourists appear to visit industrial sites relatively often, probably because they are used to this kind of tourism. Visitor statistics of the event Made in Angers demonstrate that older people (50+) are overrepresented in visitor figures, which is a general conclusion we also found in the other cities.

- Also in **Turin** we found that company visits under the umbrella of Made in Torino are particularly popular among senior citizens and visitors. Remarkably, young people with an age up to 16 cannot participate because of safety restrictions, while companies are generally very much interested in reaching this target group. More in general we conclude that Made in Torino attracts many visitors that are not targeted by the companies. With the programme, the tourism organisation Turismo Torino deliberately also targets tourists (people who stay in the city for a long weekend), but because the capacity is limited many tourists cannot participate. The companies we analysed are generally more selective in the acquisition of visitors than the tourism office, although pen producer Aurora is clearly an example of a firm that has adopted the philosophy that any contact with the outside world can be valuable.
- In **Shanghai** we observed that the Shanghai Industrial Tourism Promotion Centre targets a very broad audience, but with a specific focus on business groups (excursions, business trips) and the Japanese market (apart from the Chinese market of course). Also groups of leisure tourists (of all ages) are targeted; individual tourists are less relevant because most Chinese tend to travel in groups. Travel agencies take care of organising visits to the firms, and the impression is that they are not very selective. What we *have* seen, however, is that a firm like Shanghai Volkswagen takes care of organising visits for groups that are more important for the firm's business, just like the big companies in St Nazaire (Pays de la Loire) do.
- In **Rotterdam** industrial tourism mainly targets business visitors and educational visitors; only during the World Port Days also individual citizens and tourists can visit the firms. We found, however, that notably companies in the port and industrial complex are increasingly concerned about shortages on the labour market, implying that students, but also families with children (all potential future employees) gain importance as target groups. While some companies open their doors to primary stakeholders only (e.g., Verstegen and Smit Internationale), other firms such as the Happy Shrimp Farm and Schmidt Zeevis seem to have adopted a

more proactive approach, arguing that secondary stakeholders (the general public, citizens, the media, politicians, etc.) are important as well for the long-term continuity (also in view of corporate citizenship).

Overall, we conclude that many groups can be targeted and reached with industrial tourism. In general, public bodies define target groups broader than companies do. For instance, tourism organisations that aim to increase the number of bed nights, are more interested in visitor numbers and maybe not that much in what kind of visitors arrive. Firms tend to focus on primary stakeholders (customers, suppliers, students, etc.) and in many cases exclude secondary stakeholders (citizens, tourists, etc.) from visiting their site. In other cases, they do receive secondary stakeholders, but emphasise that they are not their target group. Many companies that open their doors to visitors without a business or educational motive report high shares of seniors (50+), in most cases not exactly the most relevant group, particularly not in view of attracting future employees.

Companies that are willing to receive citizens and tourists are mostly consumer-oriented firms, but in that case these groups can be considered (potential) primary stakeholders (buyers of their product). Also firms that are (partly) owned by the government – such as port and transport authorities in Rotterdam, Cologne and St Nazaire, and several companies in Shanghai – open their doors to the general public but that is not only the result of an economic weigh-off but presumably also the result of giving account to the tax payers who are indirectly shareholders of these firms. More and more companies without government involvement, however, also realise that it is in their interest to build sustainable relations with all actors that give them a license to operate. In this 'open door policy' firms still target specific groups but without closing the door for any group.

Looking at the geographic reach of industrial tourism (scale), we conclude that regional target groups are most relevant, particularly if the ambition is to change the image of the (manufacturing) industry as employer and responsible citizen. Particularly open door event such as the World Port Days, Made in Angers and Expedition Colonia seem to target a regional or national audience. Made in Torino, which is in fact also an event, is an interesting exception because the initiators want to use company visits as a tool to enrich Turin's image abroad, demonstrating that the city is more than Fiat. If the ambition is to attract many foreign tourists we would recommend focusing on specific countries where people are used to visiting companies: Great Britain, the US, China and Japan would probably meet that condition.

5 How to make industrial tourism attractive and accessible?

The attractiveness of an industrial tourism product depends on the company and its products: firms that produce goods with a symbolic character for the region, branded goods, consumer goods, goods of everyday life, luxurious goods, technologically demanding goods, special interest goods and/or handicraft goods are probably in a better starting position than companies that produce other kinds of goods. Taking this into account, however, company visits can be made more attractive by improving the quality of visitor services (e.g., through investments in facilities and professional guides) and making them better accessible. The case studies give several suggestions on how to do this:

- The most important conclusion in **Wolfsburg** is that the combination of a theme park and a factory visit make industrial tourism very attractive, as it combines leisure and learning (with a museum, interactive games and exhibitions) with 'the real thing'. For the factory visits, Volkswagen only uses professionally trained guides with work floor experience, and only allows visitors to visit their sites if machines are running, for the sake of authenticity. The five-star Ritz Carlton hotel, with an outdoor swimming pool in front of the factory and the power plant, is a spectacular element of Autostadt. Another strong feature is the landscape which demonstrates that an industrial site can be made very appealing. The accessibility of the theme park and the factory has been secured through the construction of a bridge that connects the site with the high-speed train station.
- In **Cologne** we learned that Bayer's communication centre benefits from its location: an attractive park landscape with a beautiful Japanese garden, easily accessible by public transport. The Rheinauhafen, currently redeveloped by the port and transport authority, partly owes its popularity as a destination to its favourable location near the city centre. In general, we conclude that companies with a less favourable location (outside the city centre, difficult to access by public transport) find it difficult to attract visitors. Furthermore we observed that companies aim to secure the quality of tours by handling a maximum group size in order to enable sufficient interaction between the guide and the participants. The Cologne case study also shows that company tours can be made attractive by arranging tours with prominent guides (Expedition Colonia) or organising tours at night (UPS). Another method is to invite students to come up with innovative ideas to improve the accessibility and attractiveness of the industrial tourism product, as in the case of industrial park Knapsack.
- The association Visit Our Companies (**Pays de la Loire**) plays a key role in making sure that company visits meet certain quality standards, concerning the availability of visitor facilities, the provision of information on the website and the skills of guides. Several obligations apply to members, acting as an incentive for other companies to meet the conditions too. In addition, members can attend workshops and courses on how to make their product more attractive. In St Nazaire the tourism office of Escal'Atlantic has invested in visitor facilities such as an elevated walkway, information displays and signing to make tours more attractive and accessible. Cointreau (Angers) is a good example of a company that succeeded to turn a company

visit into an experience by building a museum and a trendy bar right next to the distillery.

- The case of **Turin** shows how companies make a tour attractive by small changes in their production schedule to show the audience some action, as design company Gufram reported (although admittedly, these tours only take place two or three times a year). Pen producer Aurora made its product more appealing by placing art objects and information displays in the workshops, and turning one room into a photo gallery with famous people who visited the site too. For the Made in Torino tour several quality standards apply: a maximum group size, professional guides and a maximum distance between the city centre (where tourists are picked up by bus) and the site to be visited.
- The **Shanghai** Industrial Tourism Promotion Centre has developed quality standards that participating companies have to meet, and provides firms advice on how they can. The Annual Ticket Book, a voucher book developed by the centre, makes company tours more accessible. Yakult, a Japanese company, has made it tours attractive and accessible for local citizens by investing in a closed glass walkway through the factory (a company policy for all new factories!), arranging shuttle buses to the company and developing special educational programmes for children and older people. Furthermore we found that travel agencies combine factory tours with visits to other, more traditional, tourist attractions nearby.
- **Rotterdam** expects to benefit from the opening of the Rotterdam Port Experience: a theme park for the port and industrial complex, and a joint portal for actual visits to the port (by boat, bus or bike) and the port-related companies. Fast connections between the port and the city – notably by boat – are expected to further enhance the accessibility of the industrial tourism product. The experiences of three companies in the food sector demonstrate that so-called 'barriers' for industrial tourism (hygiene, security, etc.) can be turned into attractive and exciting features of the product. We also conclude that particularly non-consumer oriented firms are able to attract visitors because of innovative production and management methods, with the Happy Shrimp Farm as one of the perfect examples.

Most suggestions mentioned above can easily be applied to any company, although some ideas clearly require more investments than other ones: elevated walkways, bridges and museums are relatively expensive. Quality standards can be very effective, notably if supported by regional and national governments as in the case of Pays de la Loire and Shanghai. National or regional 'models of excellence' can stimulate other companies to improve their offer. It makes sense to make a distinction between 'basic conditions' such as parking facilities, safety procedures, a maximum group size and visitor toilets (that every company should meet) and conditions that should be preferably met to bring products on a higher quality level, such as multilingual guides and a room for video presentations.

Elevated walkways and closed glass walkways can be particularly helpful to reduce conflicts between industrial tourism and the day-to-day production, also in view of hygiene and safety standards. It is cheaper to include these facilities in an original construction than to modify a building afterwards.

One of the most interesting conclusions is that combinations can make industrial tourism more attractive. Packages may include visits to several companies as in the case of St Nazaire (Aker Yards and Airbus). In practice, however, it seems that visitors are more interested in combinations between industrial tourism and other, more traditional, tourism attractions. Also in Rotterdam, the foundation Industrial Tourism gets many requests to develop complete packages in which company visits are just one element. This implies that in terms of promotion, it is advisable to promote company visits not only as an integrated part of 'economic discoveries' (with industrial heritage, industrial museums, etc.) but also as one of the ingredients of the total tourism supply.

6 Benefits in terms of income and image for city and enterprise

Company visits generate numerous benefits for firms as well as for the cities in which they are located. In this section we discuss the benefits in terms of income and image for city and enterprise, and the convergence of interests between public and private sectors.

What are costs and benefits for companies?

In our research framework we argued that industrial tourism can be an adequate and cost-efficient instrument for marketing (promoting the firm and its brands), public relations (building sustainable relations with society, giving evidence of corporate citizenship), human resources management (attracting and motivating employees) and business development (generating direct and indirect income). Of course these benefits have to outweigh the costs of opening the firm's doors to visitors, such as the required investments in visitor facilities, and safety and security in particular. How does this work out in the case studies?

- Volkswagen (**Wolfsburg**) considers factory visits, free of charge, as a cost-efficient and effective tool for promoting the firm and its brands. An internal study has proven that these visits help to improve the company's image and to raise customer loyalty, which in the end leads to higher sales and profits. Similar motives explain the development of Autostadt, since only 70 per cent of the costs are covered by entrance fees.
- In **Cologne** the companies we analysed have opened their doors to visitors for free not only because they expect benefits in terms of human resources management (e.g., Knapsack, Bayer), but also to build good relations with customers and other stakeholders (e.g., Bayer, Rheinenergie) and to improve

the overall image, also giving evidence of corporate social responsibility (e.g., Bayer).

- The selected companies in **Pays de la Loire** mostly participate in industrial tourism because they expect their image and reputation to improve (also enhancing the relation with the municipality and society in general) as well as better access to qualified labour (Cointreau being an exception; they mainly use company visits to promote the brand). All companies we analysed charge entrance fees for the general public, mainly to cover the costs of professional guides and visitor facilities.
- The three firms in **Turin** that we investigated recognise industrial tourism as important instruments of public relations and marketing. Interestingly, the main motive for Gufram to participate in Made in Torino is showing good citizenship. For Aurora, company tours are used to emphasise that their products are 'made in Italy', which enables the firm to sell its products in higher segments of the market.
- In **Shanghai** most companies we interviewed (SVW, Yakult, companies at M50) are involved in industrial tourism to promote the company and its brands, expecting (short-term) increases in sales (even on the site, as in the case of M50). Yakult considers industrial tourism more effective and cost-efficient than traditional advertising methods. Baosteel sees industrial tourism also as a method to demonstrate that the firm pays attention to its social and environmental responsibilities.
- The companies that we analysed in **Rotterdam** expect benefits from company visits in terms of reputation and relation management, though with different views on what relations have to be managed. Many firms in the port and industrial complex open their doors to gain access to qualified labour. Schmidt Zeevis expects benefits from all communication with the outside world, and expects company visits to make employees feel proud about their work. The Happy Shrimp Farm makes a direct profit on company visits for business groups, which is partly used to subsidise visits by educational groups.

In general we can conclude that companies expect substantial benefits of industrial tourism in terms of income and image, but also in terms of relation management (including motivating employees). In addition, the case studies demonstrate that consumer-oriented companies are more interested in promoting their brands, while business-to-business companies tend to be triggered by their interest in a good relation with society and the need for well-qualified labour. In some individual examples we have observed that company visits generate direct income, in terms of ticket sales or product sales, but in most cases these revenues are too small to justify the investments and the associated costs. In most cases the income from ticket sales only covers the costs of professional guides. This implies that indirect and intangible benefits are larger and more relevant than direct, tangible benefits. The case studies also confirm that the increasing pressure on the license to operate – companies have to be good citizens – makes companies aware of the need to open their doors to society.

What are costs and benefits for cities?

For cities, industrial tourism can generate benefits in terms of additional income (attracting more visitors without major investments and using existing attractions to the full), and in terms of a more realistic image (closing the 'perception gap'). What are the main observations in the case studies?

- In **Wolfsburg**, Autostadt and the Volkswagen factory generate substantial numbers of business visitors, excursionists and tourists, but most income effects are generated within the industrial theme park which has its own hotel and catering facilities (it functions more or less as an island in the city). More importantly, however, Autostadt has also put Wolfsburg on the mental map of tourists, making it possible to develop new tourist attractions such as the Phaeno Science Centre (2005) and a designer outlet centre (2007) that may generate more spill-over effects for the local economy. The City has no share in the costs of industrial tourism, apart from investments in public space and infrastructure.
- **Cologne** prefers to avoid the term 'industrial tourism' because it is associated with the manufacturing industry, which does not exactly fit in the city's image campaign 'Cologne is a feeling'. Nevertheless, the city does expect 'company visits' to generate positive effects, notably when it comes to promoting Cologne as Germany's media capital, showing the successful redevelopment of waterfront areas and attracting inward investment (mentioned by the Chamber of Commerce). Although company visits generate relatively small numbers of visitors, it is considered an interesting niche market, with the event Expedition Colonia (lasting three weeks) acting as a crowd puller in the off season (April). The costs for the City are relatively low.
- **Pays de la Loire** considers industrial tourism – and 'economic discoveries' in general – as a significant segment of the tourism industry, with demonstrated growth potential. Company visits not only generate benefits for the hotel and catering industry, and the retail sector, but also add to a positive and well-balanced image of the region, notably as an attractive alternative location for Paris. Local authorities promote industrial tourism because it is expected to improve the image of their city (e.g., St Nazaire) and because industrial tourism helps to create better linkages between government, business and knowledge institutions (e.g., Angers). Governments on several levels (local, regional and national) invest substantial amounts of money in the development of industrial tourism.
- In **Turin** the development of industrial tourism fits in a long-term strategy which aims to differentiate the Turin economy and improve the city's image.

Company visits are promoted to show citizens (in their role as ambassadors) and visitors the diversity and strength of Turin's manufacturing industry. As for many other developments in the region, the Olympic Winter Games of 2006 have acted as a catalyst for the development of industrial tourism. With a very limited number of participants, the income effects of Made in Torino (with an annual budget of €100,000) are very small, but the image impact is considerable also due to media attention. From a regional perspective, industrial tourism is not considered very relevant because it hardly generates bed nights, which is generally the main objective for regional tourism bodies. Moreover some discussion partners expressed their opinion that the association with 'industry' is not very desirable for tourism promotion, particularly not because the region has much more to offer.

- In **Shanghai** it became clear that industrial tourism fits in the strategy of 'the government' (on various levels) to demonstrate the performance of the economy – not only branding the city or region, but also China as a whole – paying specific attention to corporate citizenship (as in the case of Baosteel, a state-owned company). Shanghai considers industrial tourism (in the broadest sense of the word) as a key segment of the tourism industry, with substantial impacts on the economy. Moreover company visits are expected to help attracting inward investment, showing that Shanghai is not only strong in manufacturing but other sectors as well. Although we have no figures about the costs of industrial tourism, the fact that the city has set up an industrial tourism promotion centre suggests that considerable investments are made.
- **Rotterdam** expects that company visits to the port and industrial complex help to promote the companies involved as employers, also strengthening the link between companies and educational institutions. Industrial tourism may also help to further improve the image of the city in general and attract additional visitors, but strategies to take advantage of these opportunities and associated investments on behalf of the municipality seem to be lacking. The recently opened theme park Rotterdam Port Experience, partly subsidised by the municipality, might act as a catalyst for the development of industrial tourism.

Concerning the (expected) benefits of industrial tourism for cities and regions we observe interesting differences between the case studies. Shanghai and Pays de la Loire see industrial tourism – in the broader definition – as a significant and growing segment of the tourism industry with positive contributions to place branding. This approach is obviously related to the supply of industrial tourism products, but it also reflects the position of these regions *vis-à-vis* other regions in their countries (China and France). Both regions have to compete against cities such as Paris and Beijing, both endowed with more traditional tourist attractions. Hence, industrial tourism is part of their positioning strategy.

In Turin and Cologne the expected benefits are smaller: industrial tourism is considered a small niche, but it can play a role in place branding though not necessarily towards tourists who might not appreciate the association with industry. Particularly for Turin we observed differences between the city and region. While the city seems to consider industrial tourism as part of a positioning strategy (in competition with cities such as Venice and Rome), the region has clearly identified other key sectors, which makes sense in view of the limited contribution of industrial tourism to the number of bed nights. Also in Cologne, which is blessed with many traditional tourist attractions, they have adopted this latter view.

Rotterdam, Turin and Wolfsburg have in common that their economy and image are strong linked to one sector: the role of the port in Rotterdam can to some extent be compared with the roles Fiat and Volkswagen play in the two other cities. For this type of cities, industrial tourism can be an instrument to demonstrate the diversity of their economies and supply of company visits. Since these key sectors will never disappear from people's perceptions, it could make sense, however, to use these companies in the positioning strategy. The tagline 'Rotterdam World Port – World City' basically says that the city is connected to the port, but also that the city has more to offer. This strategy can be translated to industrial tourism, as demonstrated by Turin: not only promoting company visits to Fiat, but also to other companies (design, food, aerospace, etc.).

It has not been our intention to make a full cost-benefit analysis of investments in industrial tourism. The case studies also explain why it is not very realistic to have this ambition. We haven't found any data that demonstrate benefits for local economies, only some indications that industrial tourism generates income (e.g., substantial shares in the visitor figures), or helps to improve the image of cities and regions. The lack of these data, however, is not specific for industrial tourism, but rather for any investment in tourism or (economic) promotion. What can be said, in general, is that the more benefits expected by cities or regions, the more costs are involved in the investments made in industrial tourism to realise these benefits. As we will see in the next section, this is also a matter of ambition and strategy.

Are there converging interests? How to benefit all?

City and enterprise have converging interests in industrial tourism: they both have an interest in a better image of the city and its industries. While companies increasingly need good relations with the communities in which they operate, public bodies more and more realise that they need the involvement of companies to reach their objectives. Although public and private interests tend to converge, companies and public actors may disagree on several issues such as the selection of target groups. This raises the question how to develop industrial tourism in such a way that it benefits all. What lessons can be learnt from the case studies in this respect?

- In **Wolfsburg** we found that the bridge between Autostadt and the city centre could be interpreted as an investment that promotes spill-over effects (to the benefit of the local economy), but this effect has not been demonstrated yet.
- The **Cologne** case study makes clear that a regional approach could be helpful to benefit all actors involved. Not only tourists but also international investors seem to look at metropolitan regions rather than at cities. Public actors in Cologne seem to understand that also visits to companies outside the municipal border (e.g., Bayer) generate benefits in terms of income and image. Also excursionists who spend their nights elsewhere in the region are appreciated.
- The most important lesson from **Pays de la Loire** is that the interests of all actors are best served if the selection of specific target groups is handed over to companies and travel agencies (also see next section). The reason is that the problem of different target groups cannot be solved on a regional level. Regional organisations should focus on joint promotion, quality improvement, and the development of networks involving companies, educational and knowledge institutions, governments, and citizen groups.
- One of the lessons from **Turin** is that the development of industrial tourism requires more sophisticated definitions of the bottom line, not only for companies but for public organisations as well. For both can be said that if they only look at direct profits (either in terms of sales or bed nights), it is more difficult to reach an agreement. When companies hold the view that every contact with the outside world can be valuable (as in the case of Aurora), it is easier to let interests converge.
- The analysis of industrial tourism in **Shanghai** demonstrates that it is easier to develop company tours to government-owned companies, because they also serve public interests. To secure sufficient supply of industrial tourism products, it is recommendable to involve these companies in any effort to develop industrial tourism. They can be an example for other firms.
- The case of **Rotterdam** shows that (expected) shortages in the labour market and a common policy agenda, in this case on how to anticipate climate change (the Rotterdam Climate Initiative), may act as catalysts for the development of industrial tourism to the benefit of all. Also in Rotterdam we came to the conclusion that converging interests can be realised sooner if companies accept the hypothesis that secondary stakeholders are as important as primary stakeholders.

The case studies produce two lessons that may apply to all cities: investments in accessibility can facilitate spill-over effects and it is advisable to develop industrial tourism on the level of the relevant functional urban region. The other remarks all concern the conflicts that may arise concerning the selection of target groups. While from a city point of view it makes sense to attract a broad audience (including leisure tourists), companies tend to define their target groups narrower, focussing on their *primary* stakeholders (business tourists). Of course, we would certainly advise cities to follow Shanghai's strategy to include public companies in industrial tourism programmes. But also the private sector will have to follow. We expect that companies become more aware of their interest in building sustainable relations with *secondary* stakeholders too, because of relations between stakeholders (a tourist can be related to a potential business partner) and the fact that one person may fulfil several stakeholder roles (the tourist of today can be a business relation tomorrow). Several examples, in particular from the cases of Rotterdam and Turin, confirm this tendency. Furthermore we think that the distinction between business and leisure tourists is becoming less relevant as both groups are looking for the same combination of content and experience (fun), though with different proportions. This implies that industrial tourism attractions can easily attract both groups, if capacity is no issue and the benefits outweigh the costs for participating actors.

7 The organisational set-up: how to organise industrial tourism?

The development of industrial tourism requires organising capacity, which essentially refers to the ability to create partnerships between actors from various sectors as well as the ability to create an environment in which such partnerships develop. What are the main observations from the case studies?

- Volkswagen (**Wolfsburg**) has its own visitor services department – employing about 30 people – with sub-units for specific target groups, enabling the company to develop tailor-made tours and to pay more than average attention to important guests (special visits and press). The board of Volkswagen gives the visitor services department relatively much freedom without claiming short-term results. Interestingly, the department also gives advice – not for free though – to other companies in the region willing to open their doors, thus helping to create an environment in which industrial tourism can develop (with or without partnerships).
- In **Cologne** industrial tourism is an insignificant side activity for both the Tourist Board and the Chamber of Commerce. We even found that the list with companies that open their doors provided by the Tourist Board is not kept up-to-date. The organisation of industrial tourism is completely in the hands of individual firms that sometimes choose to outsource part of the visits to external parties, as in the case of HGK (visits to Rheinauhafen). An exception is the business park Knapsack where the park management organisation coordinates visits.
- The association Visit Our Companies (**Pays de la Loire**) plays a key role in creating an environment that stimulates companies to open their doors, and to develop an attractive industrial tourism product. This association is financially supported by regional and national funds, but also through membership fees

(up to €500 a year) paid by the companies that also invest management time via active participation in the board of managers. In Angers, the university has taken its responsibility by developing a bachelor's programme on 'economic discoveries' in cooperation with companies and the tourism industry. In St Nazaire, the three biggest companies have opted for a model in which the local tourism bureau (for 70 per cent owned by public shareholders) arranges visits for the general public and invests in visitor facilities on their sites, while the companies take care of receiving professional visitors.

- In **Turin** we noted that the Chamber of Commerce and the tourism organisation for the Province of Turin have initiated a partnership in which they both take an active share. The industrial tourism programme Made in Torino clearly fits in a long-term strategy which makes it easier to find support among policy-makers and firms. Also worth mentioning is the strategy to expand the programme step by step – adding one or two sectors each year – which enables the initiators to convince new participants by showing the results so far, while spreading the organisational efforts (and costs) over several years. In the organisation of Made in Torino companies play a very limited role, however. Apart from their participation in the programme, companies arrange tours for more specific target groups, usually through their marketing and PR departments.
- The **Shanghai** Industrial Tourism Promotion Centre is a typical example of a coordinating body in a Chinese context: created by the local government with support of the State and in response to national policies. After its foundation the Centre started to cooperate with companies that already opened their doors, followed by a strategy to involve companies that represent the identity of the region and for that reason, are often willing to invest in their relation with the community. Important actors in the case of Shanghai are the travel agencies that organise company tours, notably for leisure-oriented visitors. Several companies (e.g., SVW) have outsourced the organisation and promotion of industrial tours for leisure visitors to one agency (while keeping the responsibility for receiving primary stakeholders within the organisational structure of the company), and in some case (e.g., Baosteel) this agency is owned by the company. For Yakult, industrial tourism plays a key role in the company's marketing strategy, explaining why the reception of visitors is a task of the public relations department. Interestingly, business park managers can also make it compulsory for companies to open their doors, as in the case of M50.
- In **Rotterdam** many organisations are involved in the development and promotion of industrial tourism, but without any explicit coordinating strategy or vision. Visits to the port and industrial complex are provided by several competing organisations, while there is an undersupply of visits to companies in other sectors. We also found that for companies, the way industrial tourism is organised depends on the objectives (e.g., recruitment, public relations, etc.).

In general we conclude that the organisation of industrial tourism requires the involvement of companies and other actors that have a stake. Ideally firms not only open their doors, but also get involved in the development and management of the regional industrial tourism product, as we observed in Pays de la Loire. Other important stakeholders to cooperate with are educational institutions (universities, schools, etc.), research institutes, travel agencies, chambers of commerce, tourism offices, the media and higher-level governments. The research demonstrates that the organisation of industrial tourism – on company and regional level – very much depends on the ambitions and aims of the actors involved: it is important to develop a vision and strategy in which the main objectives of industrial tourism (e.g., improving the image, attracting visitors, creating networks) are specified. Such a vision and strategy may help cities and regions to gain support not only from local stakeholders, but also from higher-level governments.

8 Recommendations

Our international comparative research has resulted in several lessons for cities that want to develop industrial tourism. The following eight recommendations should not be seen as a blueprint for how to develop industrial tourism, but rather as a list of helpful guidelines and valuable insights for a debate between stakeholders.

1) **Agree on the main objectives of developing industrial tourism.** Is it the aim to generate income, to improve the image of the city and its industries or to strengthen the relation between firms and society? Or are all objectives equally relevant? Some relevant circumstances to take into consideration are:
    a) The mix between consumer and non-consumer oriented firms has implications for the ability to attract leisure-oriented visitors.
    b) The relative importance of industrial tourism in comparison with other segments of the tourist industry. Particularly industrial cities without a tradition in tourism are recommended to consider including industrial tourism development in their strategies.
    c) Companies are increasingly concerned about their image and reputation in general, and as employers in particular. From their point of view, the main objectives of industrial tourism should be related to improving perceptions and stakeholder relations.

2) **Define 'industrial tourism'.** Define the borders of industrial tourism and choose the right 'name' taking into account the perceptions, expectations, needs and demands of potential target groups.
    a) The case studies make clear that the term 'industrial tourism' should at least be topic of debate, because of its association with

the manufacturing industry and leisure tourism, as well as with industrial heritage. The term 'company visits' is more neutral, but maybe also less appealing. The challenge is to find a more appealing name with the right associations: good examples are Visit Our Companies in Pays de la Loire and Made in Torino; Tour the Excellent.

b) In the definition of industrial tourism (or whatever term is used) the needs and demands of relevant target groups should be leading: their demand for 'discovering firms and the local economy' can also be met by developing communication centres or theme parks, such as Autostadt. Such facilities will never fully replace real company visits but they are very helpful to avoid conflicts between industrial tourism and the daily production process. Cities should consider including such 'showcases' in their definition of industrial tourism.

3) **Build on strong assets.** Industrial tourism can be used to enrich the image of a city. Cities are advised to open the doors of the sectors in which they are specialised as well as sectors with growh potential. In the knowledge economy the challenge for cities is to demonstrate their excellence and their ability to innovate.

4) **Identify target groups** that both city and enterprise want to receive. If the ambition is to use industrial tourism for more than attracting tourists only, companies should be involved in selecting target groups.

5) **Use industrial tourism as unique selling point for attracting tourists and lengthening their stay.** Although companies may not have a direct interest in receiving leisure tourists, they might still be convinced to open their doors for them, either regularly or occasionally. Industrial tourism events and programmes as well as jointly developed visitor centres offer possibilities to benefit both city and enterprise.

6) **Introduce quality standards for company visits.** Quality standards not only secure the attractiveness and accessibility of the visits, but also stimulate and help companies to open their doors. Not only local but also national governments may play a role in the realisation of such standards.

7) **Develop an internet portal for industrial tourism.** Make sure that all information on company tours is findable even if restrictions apply.

8) **Cooperate with travel agencies, tourism offices and tour operators.** Small and medium-sized companies but also big firms lack the human

resources and also the expertise to organise company visits: it is not their core business. This implies that specialised travel agencies, tourism offices and tour operators should play a key role in meeting the growing demand.

To conclude, our international comparative research makes clear that industrial tourism can produce benefits in terms of income and image for cities and their industries. The case studies demonstrate that a more positive and proactive approach, with the support of public and private actors, can contribute to the competitiveness of individual companies and the city in which they are located. Many cities have a good potential, but now it is up to the stakeholders to take advantage of these opportunities.

Index

Note: numbers in brackets preceded by *n* are references to footnotes.